02-CTS-7P

PROTOCOLS TO ASSESS THE BREAKDOWN OF FLUSHABLE CONSUMER PRODUCTS

by:

The Procter & Gamble Company
Winton Hill Technical Center

AF631510

2003

Water Environment Research Foundation

The Water Environment Research Foundation, a not-for-profit organization, funds and manages water quality research for its subscribers through a diverse public-private partnership between municipal utilities, corporations, academia, industry, and the federal government. WERF subscribers include municipal and regional water and wastewater utilities, industrial corporations, environmental engineering firms, and others that share a commitment to cost-effective water quality solutions. WERF is dedicated to advancing science and technology addressing water quality issues as they impact water resources, the atmosphere, the lands, and quality of life.

For more information, contact:
Water Environment Research Foundation
635 Slaters Lane
Alexandria, VA 22314-1994
Tel: (703) 684-2470
Fax: (703) 299-0742
www.werf.org
werf@werf.org

This report was co-published by the following organization. For non-subscriber sales information, contact:

IWA Publishing
Alliance House, 12 Caxton Street
London SW1H 0QS, United Kingdom
Tel: +44 (0) 20 7654 5500
Fax: +44 (0) 20 7654 5555
www.iwapublishing.com
publications@iwap.co.uk

© Copyright 2003 by the Water Environment Research Foundation. All rights reserved. Permission to copy must be obtained from the Water Environment Research Foundation.
Library of Congress Catalog Card Number: 2003106116

WERF ISBN: 1-893664-94-5 IWAP ISBN: 1-84339-675-0

This report was prepared by the organization(s) named below as an account of work sponsored by the Water Environment Research Foundation (WERF). Neither WERF, members of WERF, the organization(s) named below, nor any person acting on their behalf: (a) makes any warranty, express or implied, with respect to the use of any information, apparatus, method, or process disclosed in this report or that such use may not infringe on privately owned rights; or (b) assumes any liabilities with respect to the use of, or for damages resulting from the use of, any information, apparatus, method, or process disclosed in this report.

The Procter & Gamble Company

This document was reviewed by a panel of independent experts selected by WERF. Mention of trade names or commercial products does not constitute WERF endorsement or recommendations for use. Similarly, omission of products or trade names indicates nothing concerning WERF's positions regarding product effectiveness or applicability.

Water Environment Research Foundation

ACKNOWLEDGMENTS

Many have contributed to the development of this document. The technical leader was Drew C. McAvoy, Ph.D. (Principal Scientist). Technical advisors to the project were Thomas W. Federle, Ph.D. (Research Fellow) and Jay W. Gooch, Ph.D. (Technical External Relations Manager). Those that contributed to the development of the test methods were Gayle A. Rece (Principal Researcher), Erin L. Schwab, B.S. (Senior Researcher), Barbara A. Nuck, B.S. (Principal Researcher), and Nina R. Itrich, B.S. (Principal Researcher). Quality Assurance guidance was by Rodney C. Stark, B.S. (Senior Scientist). Others that contributed to the overall program were Kenneth W. Miller, Ph.D. (Associate Director), Pam Kloepper-Sams, Ph.D. (Section Head), Elizabeth A. (Beth) Bookland, Ph.D. (Section Head), and Burney S. Schwab, B.S. (Principal Researcher).

We would also like to thank the peer reviewers whose helpful comments significantly improved the overall approach and test methods.

Report Preparation

Principal Investigator:
Drew C. McAvoy, Ph.D.
The Procter & Gamble Company

Project Team:
Gayle A. Rece
Erin L. Schwab
Barbara A. Nuck, B.S.
Nina R. Itrich, B.S.
Rodney C. Stark, B.S.
The Procter & Gamble Company

Project Subcommittee

Damann L. Anderson, P.E., *Hazen and Sawyer*
Herschel A. (Chip) Elliott, Ph.D., P.E., *Pennsylvania State University*
C.P. Leslie Grady Jr., Ph.D., P.E., *Clemson University*
Thomas P. Konen, P.E., *Stevens Institute of Technology*

Water Environment Research Foundation Staff

Research Program Director: Jeff Moeller, P.E.
Project Manager: Mary Strawn

ABSTRACT AND BENEFITS

Abstract:

A scientifically sound approach is needed to ensure that flushable consumer products are compatible with household plumbing fixtures, as well as wastewater collection and treatment systems. In addition to assessing disposal system compatibility, an assessment approach should also ensure that flushable consumer products do not become an aesthetic nuisance in surface waters and soil environments. This document presents an overall approach for assessing the fate and compatibility of consumer products in wastewater disposal systems. While the focus of this document is on the United States, it is believed that the conceptual approach and many of the test methods could be used to assess the compatibility of flushable consumer products in wastewater disposal systems throughout the world.

Benefits:

- Comprehensive document, which describes a scientifically sound approach for assessing the fate and compatibility of flushable consumer products in wastewater disposal systems, is available for use.
- Peer review of approach ensures objectivity and removes bias from the assessment process.

Keywords: Consumer products, flushability, fate, biodegradability, compatibility, wastewater disposal systems

PREFACE

Following a program review and site evaluation visit, the Peer Review Committee concluded that the following document represents a fundamentally sound approach and provides sufficient methodological detail for the development of a successful and scientifically defensible product-testing program. While an attempt was made to develop procedures for the entire range of flushable consumer products, wastewater conveyance, and treatment systems, it is recognized that modifications or additional protocols may be necessary to address specific products, wastewater systems, and local conditions. Some procedures are still evolving and will benefit from further testing and evaluation. Thus, this document provides a launching point for others who want to refine methods and procedures or develop new ones. The document is viewed as part of an ongoing effort to ensure that consumer products are compatible with wastewater conveyance and treatment systems and that these products do not pose an environmental or aesthetic nuisance. As such, the document is viewed as a first step in developing a widely accepted and comprehensive guidance manual for assessing the compatibility of flushable products in household and waste disposal systems.

WERF Peer Review Committee
Damann L. Anderson, P.E.
Herschel A. (Chip) Elliott, Ph.D., P.E.
C.P. Leslie Grady Jr., Ph.D., P.E.
Thomas P. Konen, P.E.

WERF Peer Review Manager
Mary Strawn

TABLE OF CONTENTS

LIST OF FIGURES

Water Environment Research Foundation

CHAPTER 1.0

INTRODUCTION

1.1 Background

Several consumer products are currently being marketed as "Flushable," "Biodegradable," and/or "Safe for Sewers and Septic Systems." These products include bath tissue, wet wipes, and tampons. In the United States, the Federal Trade Commission (FTC) has the responsibility of regulating environmental claims, especially those related to existing waste management practices. It is the responsibility of the manufacturers to have supportable data for these claims. Currently, there are no broadly acceptable methodologies for assessing the flushability of consumer products. Therefore, it is necessary to have a standard approach and test methods to support the flushability of consumer products.

In developing a flushable product the manufacturer may need to balance between those properties that make a product useful and those that make it flushable. Many factors come into play in such an assessment. In particular, one must be aware of existing infrastructure for wastewater disposal. There may also be new laws that reduce flow in plumbing fixtures, affect biodegradability claims, or change the way sludge disposal is handled. Consumer perceptions are also an important factor. Changing people's habits and practices may not be easy, especially if there is an additional cost to the consumer.

Scientifically sound test methods are needed to ensure that flushable products are compatible with household plumbing fixtures, as well as wastewater collection and treatment systems. In addition to being compatible with current disposal systems, these test methods should also ensure that flushable products do not become an aesthetic nuisance in surface waters and soil environments following their collection and treatment of wastewater. In the development of an overall approach and test methodologies used for assessing flushability, several guiding principles were developed. The guiding principles presented below should be kept in mind when reading this document.

For a product to be flushable it should:

- Clear toilets and household drainlines under expected usage

- Be compatible with wastewater conveyance and treatment systems
- Not be visible in the environment within a reasonable period of time
- Be safe (non-toxic) in relevant environmental compartments

The first three principles relate to the fate and biodegradation of a consumer product and thus will be the focus of this document. The last guiding principle relates to product safety and is addressed by investigating the exposure and potential effects of the product ingredients in relevant environmental compartments (e.g., aquatic and terrestrial systems). Since the approach for conducting a safety assessment is beyond the scope of this document, it will not be presented here.

The purpose of this document is to provide guidance on an overall approach and test methods that can be used for assessing the flushability of consumer products. This assessment approach includes product compatibility in home plumbing fixtures and drainlines, as well as disposal in municipal and onsite wastewater treatment systems. While this manual represents the first comprehensive guide of methodologies for assessing the fate and biodegradability of flushable consumer products, it is hoped that subsequent improvements will be forthcoming.

1.2 Peer Review Process

At the request of The Procter & Gamble Company (P&G), the Water Environment Research Foundation (WERF) appointed a Peer Review Committee (PRC) to evaluate their approach for assessing the fate and biodegradability of flushable consumer products. The ultimate goal of this peer review was to develop a scientifically sound document that describes the approach, current methodologies, and quality assurance procedures that support the flushability of consumer products. Moreover, it was intended that this document be shared with WERF subscribers and other stakeholders so that others can benefit from these methods.

This peer review process was thought to be integral for conducting sound science. Ultimately it should lead to better data by ensuring objectivity and removing bias from the assessment process. Moreover, an independent review lends credence to the scientific approach and methods being used. Such reviews also improve the decisions being made because they will be based on best available science. Thus, this peer review was thought to be crucial in the development of sound test methods.

This peer review evaluated the overall approach, methodologies, and quality assurance procedures used to assess the flushability and biodegradability of a product. Specific questions addressed by the review panel are listed below.

1. Overall Approach: Is the overall approach for assessing the fate and biodegradation of flushable consumer products scientifically and technically sound? If not, what recommendations would be made to improve the approach?

2. Methodologies: Are the methods used for assessing the fate and biodegradation of flushable consumer products scientifically and technically sound? If not, what recommendations would be made to improve these methods?

3. Quality Assurance: Are the quality assurance procedures sufficient to support the results of the test methods? If not, what else could be done to ensure data quality?

This review process can be thought of as the first step in developing an acceptable standard approach for assessing the compatibility of flushable products in wastewater disposal systems.

1.3 Terminology

The following terminology will be used throughout this document. An understanding of these definitions should be useful to the review reader.

Biodegradation: A consumer product is considered biodegradable if its organic components are broken down into simple compounds (carbon dioxide, methane, water, etc.) by the action of naturally occurring microorganisms such as bacteria and fungi, and there is an absence of persistent metabolites or residues. For a material to be considered completely biodegradable the following must be demonstrated: 1) disappearance of the parent material, 2) substantial production of carbon dioxide (aerobic conditions) or methane (anaerobic conditions), and 3) absence of persistent metabolites.

Disintegration: The disintegration of a consumer product is the process during which it breaks up into smaller pieces by physical, chemical, or biological means. Disintegration is assessed by the weight loss of a product under specific environmental conditions.

Dispersability: Dispersability refers to a consumer product's ability to readily break apart due to physical forces.

Degradation: Degradation is used interchangeably with disintegration.

Flushability: A consumer product is considered flushable if it is able to pass through the toilet bowl and household drainline, is compatible with onsite and municipal wastewater treatment systems, and disintegrates such that it is not recognizable in the environment over a reasonable period of time.

Removal: A consumer product can be removed from wastewater by screening, settling, and/or degradation processes.

Synthetic Fecal Matter (SFM): SFM refers to a plant-derived fiber material that has similar physical characteristics as human fecal matter.

CHAPTER 2.0

FLUSHABILITY AND BIODEGRADABILITY

This chapter presents the overall approach and current test methods used for assessing the flushability and biodegradability of consumer products. The pathways of disposal are presented first, a discussion on flushability versus biodegradability second, an overview of the testing strategy third, and finally, a brief description of the quality assurance program.

2.1 Disposal Pathways of Flushable Products

Because of its convenience, the disposal of consumer products via the toilet is a common practice. Consumer products specifically designed for disposal in household plumbing are being developed to take advantage of this convenience. For a product to be flushable, it must pass through the household toilet and drainage pipe system, be transported in wastewater conveyance systems, and be compatible with wastewater treatment systems. A schematic of potential pathways for flushable consumer product disposal is presented in Figure 2-1.

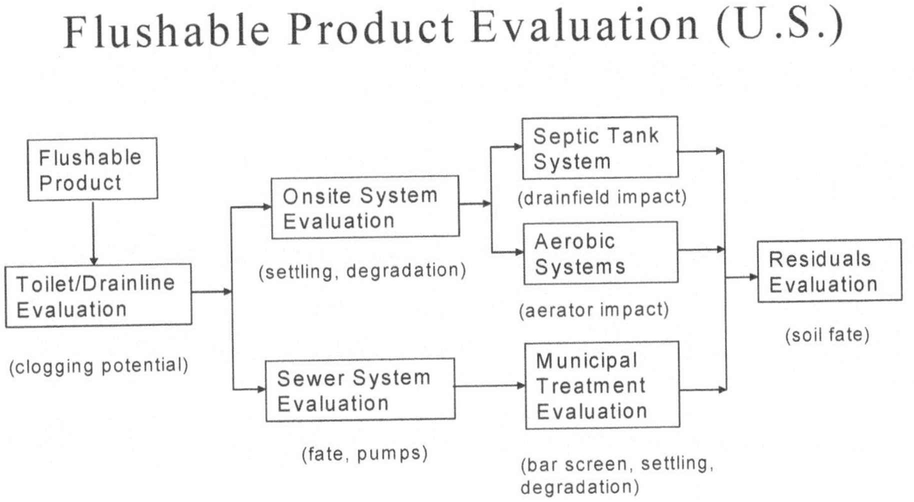

Figure 2-1. Disposal pathway of flushable consumer products in the United States

The first step in the process is to assess the product's compatibility with toilets. In particular, a flushable product should be able to pass through a toilet without clogging at anticipated usage rates. Standard test methods exist to evaluate the performance of toilets using sponges, polypropylene balls, polyethylene granules and dyes (e.g., ASTM), but none of these performance evaluations use actual consumer products. Therefore, it is necessary to develop specific test methods for assessing the acceptance of flushable products in toilets.

Most toilets in the U. S. require the establishment of a siphon in the trap before a material in the bowl can be evacuated. This design worked reasonably well in the past when toilets used between 3.5 and 7 gallons of water per flush, but when the national standards in the mid-1990s required a 1.6-gallon (6 L) flush toilet, the flushing of consumer products became more challenging. This move to a low consumption toilet has caused manufacturers to redesign their toilets. Because there are a large variety of toilets in use, the assessment of toilet clearance is challenging.

Minimum standards exist for the construction of household drainage pipe systems, which include allowable pipe diameters and slopes, venting requirements, and piping materials. The passage of solid materials through these systems is dependent on the water being able to move the product. If a product deposits in the drainage pipe, then the next passage of water must be able to resuspend and transport that material. If the product is resistant to resuspension, then the potential exists for blockage to occur in the pipe. Some households may have a sewage ejector pump to lift toilet water to a gravity drainage pipe (e.g., in a newly remodeled basement). In this situation, an evaluation of flushable products may still be warranted, even though pump manufacturers generally recommend not to flush anything other than bath tissue down the toilet.

Municipal conveyance systems typically transport products easier than home drainline system because of the greater amount of continuous flow. Products should, however, be evaluated for their compatibility with pumping lift stations. A high strength product could present a problem in these pumping systems. For example, small diameter pressure sewer systems are sometimes used in areas where the topography does not allow for the use of gravity sewer systems. In these systems, wastewater from the home is pumped using a grinder pump into a small diameter pressure pipe. An evaluation of flushable products in these systems may be warranted, even though pump manufacturers generally recommend that nothing be flushed down the toilet except bath tissue.

In municipal wastewater treatment, the removal of large solids is accomplished by screening on bar racks (1″ opening) or grinding in comminutors followed by settling in a grit chamber. Materials removed in these treatment processes are typically landfilled. Once the large solids have been removed, the next treatment process is primary sedimentation. Exceptions to this would be lagoon or oxidation ditch treatment. Primary sedimentation provides removal of solids, fats, oils, grease, and other floatable materials. Its function is to remove suspended soils prior to biological treatment. Solids removed via settling in the primary clarifier are pumped to sludge processing. The floatable material is skimmed and then landfilled. Following the secondary biological treatment process, the solids are settled in a final clarifier and some of these solids are wasted from the system where they enter the sludge processing system. Sludge produced from the primary clarifier and secondary biological process requires further treatment prior to disposal. The sludge is typically digested via aerobic or anaerobic processes. The dewatered digested sludge may then be incinerated, landfilled or land-applied.

Approximately 25% of all households in the U.S. have onsite disposal systems. Most of these systems are comprised of a septic tank with a soil absorption drainage field, but aerobic treatment systems are becoming more popular. Both onsite treatment options involve the production of sludge, which may be land-applied, land-filled or disposed of at a municipal wastewater treatment plant. The most important aspects of a flushable product when assessing its compatibility in these systems are material strength, buoyancy and biodegradability.

In a septic system, the heavier solids will settle in the septic tank, the lighter material will float to the scum layer, and the remaining liquid portion will be transported to the drainage field. Degradation of the settled solids occurs by anaerobic and facultative organisms, which converts organic materials to methane and carbon dioxide gases. The settled solids can then entrap these gases and depending on their density, can cause them to resuspend in the bulk solution. If solids escape the tank, they can potentially clog the drainage field pipe, causing premature failure of the system. Maintenance of a septic system generally involves pumping the accumulated sludge from the tank every one to five years, depending on the solids loading and the tank capacity. The material pumped from the tank, called septage, is disposed of by land application, landfilling or at a municipal wastewater treatment plant.

In aerobic onsite treatment systems, the wastewater typically passes through a small settling tank prior to being treated in an aerobic chamber. Following aerobic treatment, the wastewater is either passed through a sand filter or a soil absorption drainage field. These systems are generally a scaled-down version of a municipal treatment system. The potential impact of a product in an aerobic onsite treatment system is similar to those in a septic system. If a product is not removed in the settling chamber, then mechanical problems may occur in the biological treatment unit. In particular, if the product retains its physical integrity, then it may interfere with the performance of the aerators. Also, if the product is carried out of the aeration chamber, it may cause problems in the drainage field or sand filter.

2.2 Flushability Versus Biodegradability

Biodegradability is an attribute that can significantly contribute to consumer, societal, and regulatory acceptance of flushable products. However, it is not always essential from a scientific perspective that a flushable product be completely biodegradable. For example, in municipal wastewater treatment plants, large solids are often removed by screening with bar racks, grinding with comminutors, or settling in grit chambers. These solids are typically landfilled or incinerated. Thus, biodegradability would have little relevance for a product or its constituents if it is removed from wastewater in this manner. Assuming the product does not become a nuisance at the treatment plant by being present in sufficient quantities to clog the bar rack or comminutor, removal by screening, grinding or settling could represent an acceptable fate for a flushable product. Likewise for septic systems, if a product settles and is retained in the septic tank it would be considered compatible with the system even though it might not anaerobically biodegrade. Therefore, if a flushed product is not completely biodegradable, but is compatible with prevailing wastewater treatment systems and is removed from wastewater such that it does not represent a significant risk of becoming a recognizable component (or "trash") in the environment, it could be considered flushable.

From a scientific perspective, the potential importance of biodegradation increases as a product or its constituents pass into primary and secondary wastewater treatment systems or enter onsite wastewater treatment systems. In these situations, the biodegradation of a product or its breakdown constituents could be an important removal mechanism. Biodegradation would be particularly important for the product components that settle and end up in the sludge of these treatment processes. For municipal wastewater treatment, the sludge is generally reduced in volume through aerobic or anaerobic digestion. For onsite treatment, the sludge generally resides in an anaerobic environment. It is common in some locales, particularly in rural areas, for the sludge from these systems to be amended to soil. A product or constituent that survives these processes in an identifiable form would represent a potential nuisance. The level of this nuisance will grow if the material persists in soil and accumulates with each additional sludge amendment. Thus, the use of biodegradable materials can help mitigate the risk that a solid material will transit wastewater treatment systems to an environmental compartment or create excessive build-up in onsite treatment systems.

2.3 Assessment Approach and Methodologies

A systematic approach for assessing the compatibility with existing wastewater disposal systems, as well as the environmental fate and biodegradation of flushable consumer products is presented below. This approach is based on the natural flow of product through the conveyance and treatment processes. The testing strategy is also presented. A general description of the test methods is presented in Chapters 3.0 through 6.0, while a detailed description of the test methods is provided in the Appendix. Each test method has its own acceptance criteria, which will either support a flushable or biodegradable claim or will require additional higher tier testing. The entire set of test results is then used in a final flushability assessment. Several examples of how to use the testing strategy are presented in Chapter 7.0.

2.3.1 Flushability

Before conducting a flushability assessment, one should have a basic understanding of the composition of the product. In particular, information on a product's wet strength and/or density may be helpful in identifying the most important pathways and questions to be asked. Knowledge about the percentage of non-biodegradable constituents as well as usage rate may also be extremely useful when assessing a product. This type of information would be helpful in positioning the magnitude of a particular concern or potential issue.

When assessing the flushability of a product, one needs to answer a series of questions. These questions are depicted in a flow chart for each of the various pathways a product may take after being flushed. The question flow chart for household fixtures and drainlines is presented in Figure 2-2. The successful answering of these questions would lead one to assess the product's compatibility with municipal and/or onsite wastewater treatment. The question flow chart for assessing product compatibility with municipal wastewater treatment is presented in Figure 2-3 and the question flow chart for onsite treatment is presented in Figure 2-4.

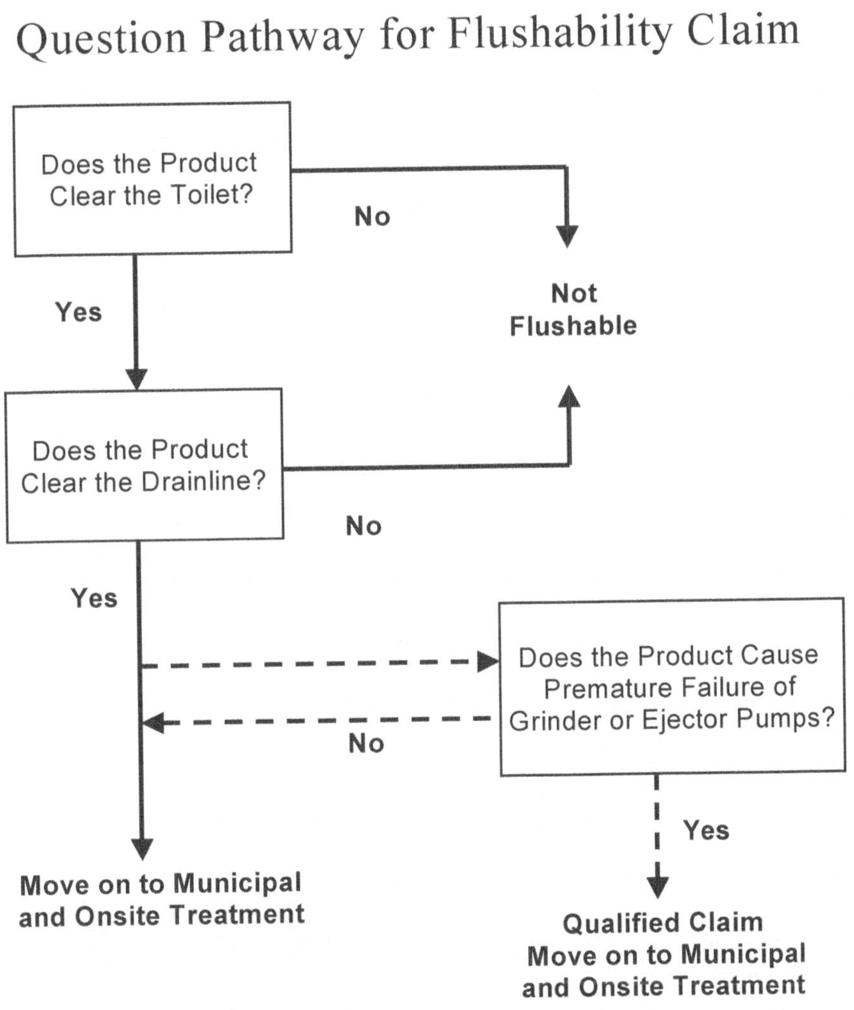

Figure 2-2. Question flow chart for assessing a product's compatibility with household plumbing fixtures and drainlines

Question Pathway for Flushability Claim

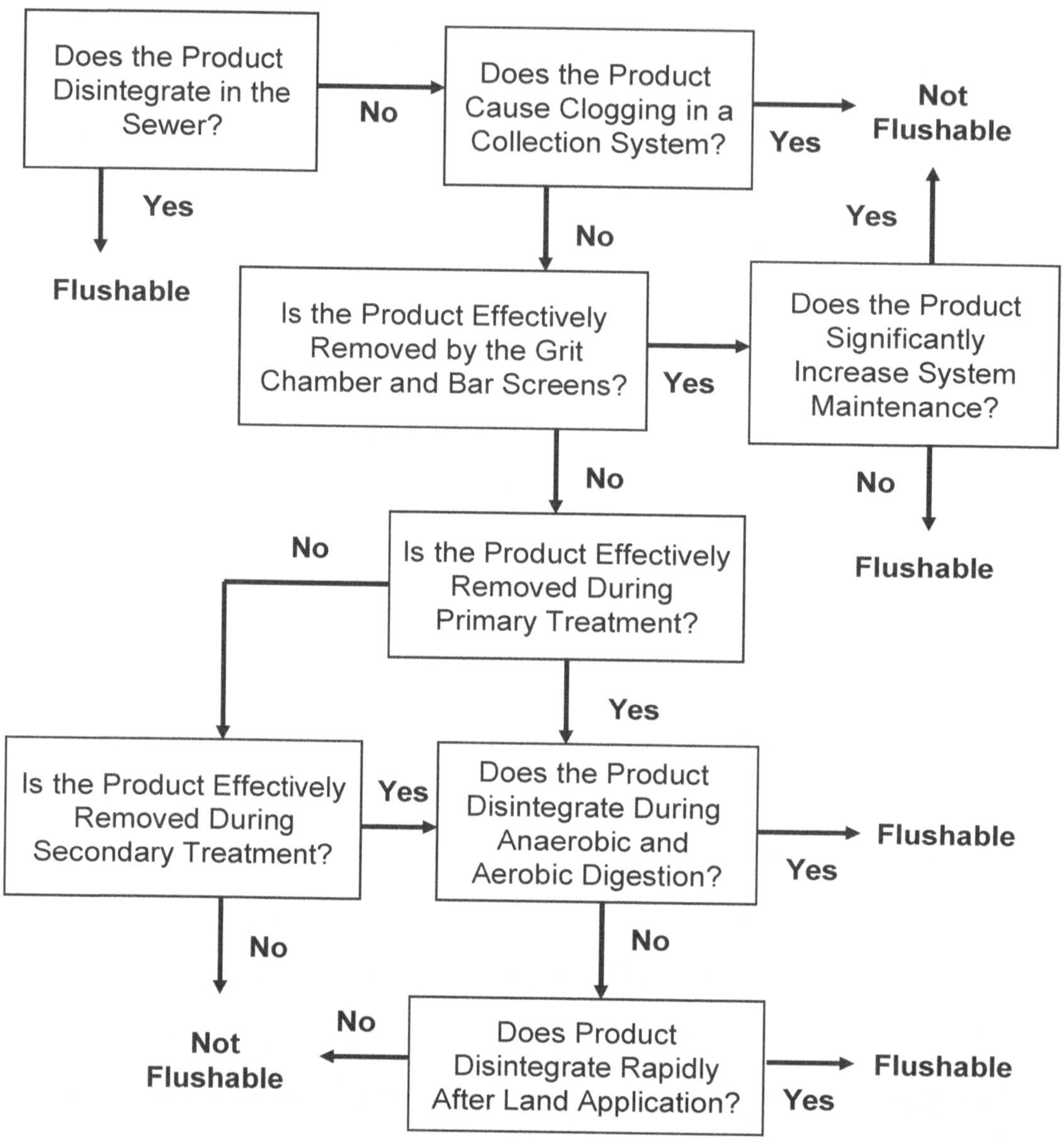

Figure 2-3. Question flow chart for assessing a product's compatibility with municipal collection and treatment systems

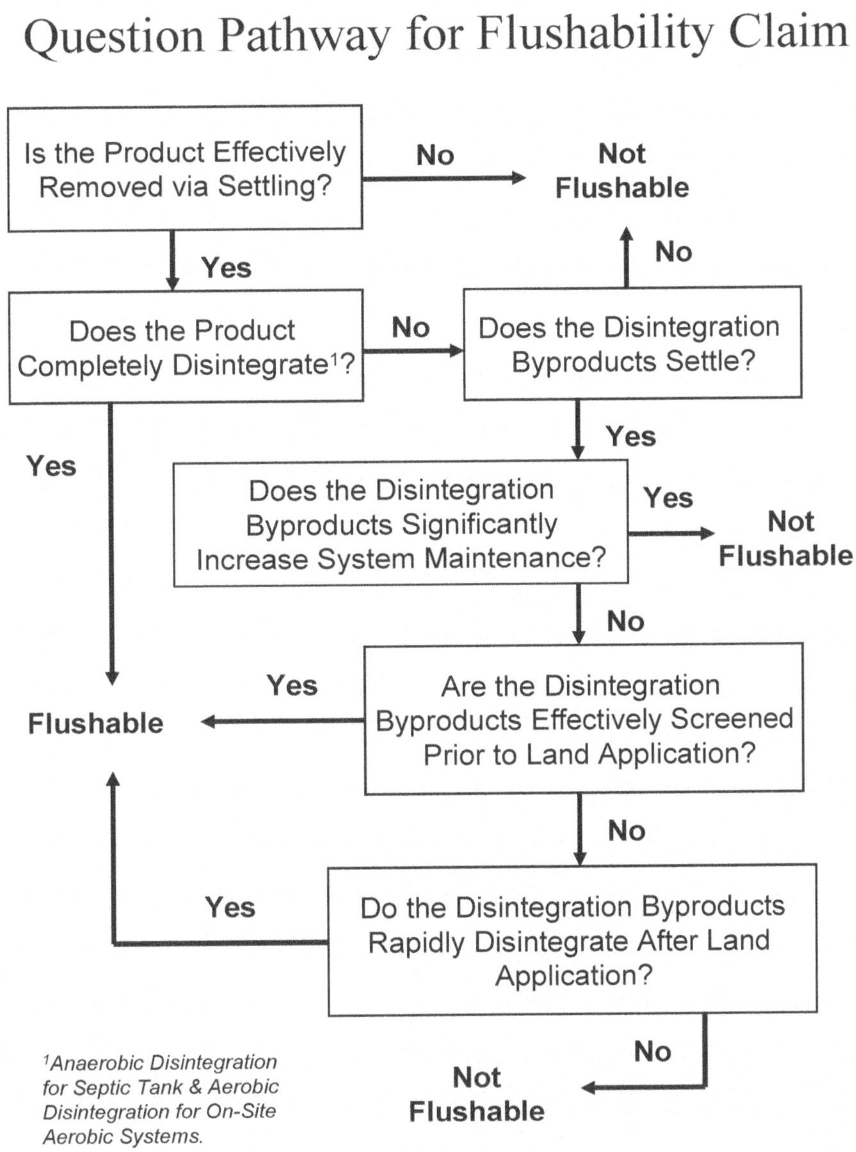

Figure 2-4. Question flow chart for assessing a product's compatibility with onsite treatment systems

The testing strategy to answer the flow chart questions for flushability is presented below. Several tests have been developed to address each question and in some cases a single test may answer several questions. These tests are set up in a tiered fashion. In principle, the initial round of testing (Tier 1) is intended to be rapid and low-cost. Because these tests may not be entirely representative of an actual system, their acceptance criteria are generally set high to account for uncertainty in their prediction. The second round of testing (Tier 2) is generally more realistic and may incorporate laboratory flow through systems. These tests generally take longer to conduct and are more costly. The third group of tests (Tier 3) is the most realistic and is typically conducted in full-scale field systems. Field tests generally take longer to conduct and require more personnel than do the first and second round of tests. Their major advantage, though, is that field tests significantly reduce the uncertainty in predicting a product's fate.

Below is the testing strategy to answer the questions presented in Figures 2-2 through 2-4. This strategy is used to generate product-specific evidence for support of flushable claims. For each question, the approach starts with rapid screening tests and, if needed, proceeds to more realistic laboratory tests, and ultimately to longer-term field tests. This strategy allows for the product developer to conduct the most appropriate amount of testing in support of a flushability claim. Note that when following through the question flow charts in Figures 2-2 through 2-4, not every question needs to be answered for a product to be flushable.

Household Conveyance:

The first set of questions deals with household plumbing fixture and drainline compatibility (Figure 2-2).

Question: Will the product clear the toilet under expected usage conditions?
Testing Strategy
Tier 1–Toilet and Drainline Screening Test
Tier 2–Toilet and Drainline Confirmatory Test
Tier 3–Home Use Toilet and Drainline Test

Question: Will the product cause abnormal and unacceptable clogging in household drainlines?
Testing Strategy
Tier 1–Toilet and Drainline Screening Test
Tier 2–Toilet and Drainline Confirmatory Test
Tier 3–Home Use Toilet and Drainline Test

A secondary issue for some households is the compatibility of a flushable product with grinder and sewage ejector pumps (Figure 2-2).

Question: Will the product cause premature failure of grinder or sewage ejector pumps?
Testing Strategy
Tier 1–Aerobic Disintegration Shake Flask Test
Tier 2–Grinder and Ejector Pump Test
Tier 3–Home Use Toilet and Drainline Test

Municipal Wastewater Collection and Treatment:

The first set of questions deal with sewer pipe conveyance (Figure 2-3).

Question: Will the product disintegrate during conveyance to a municipal wastewater treatment plant?
Testing Strategy
Tier 1–Aerobic Disintegration Shake Flask Test
Tier 2–Tipping Pipe Test (Proposed)
Tier 3–Sewer Conveyance/Municipal Treatment Test

Question: Will the product cause abnormal and unacceptable clogging in municipal collection systems?

Testing Strategy
Tier 1–Toilet and Drainline Screening Test
Tier 2–Toilet and Drainline Confirmatory Test
Tier 3–Sewer Conveyance/Municipal Treatment Test

The next set of questions deal with municipal wastewater treatment plant removal and compatibility (Figure 2-3). Note that these tests are only needed if the product reaches a wastewater treatment plant in a recognizable form.

Question: Will the product be effectively removed in the grit chamber and by bar screens?

Testing Strategy
Tier 1–Column Settling Test
Tier 1–Aerobic Disintegration Shake Flask Test
Tier 2–Tipping Pipe Test (Proposed)
Tier 3–Sewer Conveyance/Municipal Treatment Test

Question: Will the product loading rate cause a significant increase in system maintenance?

Testing Strategy
Tier 1–Estimate Loading Rate Based on Expected Product Usage
Tier 3–Sewer Conveyance/Municipal Treatment Test

Question: Will the product be effectively removed during primary treatment?

Testing Strategy
Tier 1–Column Settling Test
Tier 3–Sewer Conveyance/Municipal Treatment Test

Question: Will the product disintegrate during aerobic and anaerobic digestion?

Testing Strategy
Tier 1–Anaerobic Disintegration Test
Tier 1–Activated Sludge Disintegration Test (Proposed)
Tier 2–Porous Pot Test
Tier 3–Sewer Conveyance/Municipal Treatment Test

Onsite Wastewater Treatment:

The next set of questions deal with product compatibility in onsite wastewater treatment (Figure 2-4).

Question: Will the product or its disintegration by-products escape a septic tank and contribute to drainfield clogging?

Testing Strategy
Tier 1–Column Settling Test
Tier 1–Anaerobic Disintegration Test

Tier 3–Septic System Test

Question: Will thc pumped septage from a septic tank contain intact product or recognizable fragments that may be visible following land application?

Testing Strategy

Tier 1–Column Settling Test

Tier 1–Anaerobic Disintegration Test

Tier 3–Septic System Test

Question: Will the product loading rate cause a significant increase in septic system maintenance (pumping)?

Testing Strategy

Tier 1–Estimate Loading Rate Based on Expected Product Usage

Tier 3–Septic System Test

Question: Will the product or its disintegration by-products escape an aerobic system or cause adverse affects to its overall operation?

Testing Strategy

Tier 1–Column Settling Test

Tier 1–Activated Sludge Disintegration Test (Proposed)

Tier 2–Porous Pot Test

Tier 3–Aerobic System Test

Question: Will the pumped septage from an aerobic system contain intact product or recognizable fragments that may be visible flowing land application?

Testing Strategy

Tier 1–Column Settling Test

Tier 1–Activated Sludge Disintegration Test (Proposed)

Tier 2–Porous Pot Test

Tier 3–Aerobic System Test

Question: Will the product loading cause a significant increase in aerobic system maintenance (pumping)?

Testing Strategy

Tier 1–Estimate Loading Rate Based on Expected Product Usage

Tier 3–Aerobic System Test

Soil Amendment:

The following question deals with sludge and septage amendment to soil (Figures 2-3 and 2-4). These tests are only conducted if it is determined from previous testing that a product or recognizable fragments would be visible in the soil environment following sludge/septage application.

Question: Will the product or its disintegration by-products present in septage and sludge persist long-term after land application?

Testing Strategy

Tier 1–Aerobic Biodegradation Test

Tier 2–Laboratory Soil Disintegration Test (Proposed)
Tier 3–Field Soil Disintegration Test

2.3.2 Biodegradability

There are few absolute technical requirements for defining biodegradability as a product claim. Internal and external scientific norms revolve around *weight-of-evidence* and common sense principles. The FTC has published the following in the Green Guide:

> Degradable/biodegradable/photodegradable: It is deceptive to misrepresent, directly or by implication, that a product or package is degradable, biodegradable or photodegradable. An unqualified claim that a product or package is degradable, biodegradable or photodegradable should be substantiated by competent and reliable scientific evidence that the entire product or package will completely breakdown and return to nature, i.e., decompose into elements found in nature within a reasonably short period of time after customary disposal. Claims of degradability, biodegradability or photodegradability should be qualified to the extent necessary to avoid consumer deception about: 1) the product or package's ability to degrade in the environment where it is customarily disposed; and 2) the rate and extent of degradation.

By using an example, the FTC has also provided guidance on the relevance of a biodegradability claim to a material that is not disposed of in a manner in which biodegradability is relevant or realized.

> A trash bag is marketed as "degradable", with no qualification or other disclosure. The marketer relies on soil burial tests to show that the product will decompose in the presence of water and oxygen. The trash bags are customarily disposed of in incineration facilities or at sanitary landfills that are managed in a way that inhibits degradation by minimizing moisture and oxygen. Degradation will be irrelevant for those trash bags that are incinerated and for those disposed of in landfills, thus the marketer does not possess adequate substantiation that the bags will degrade in a reasonably short period of time in a landfill. The claim is therefore deceptive.

Based on the FTC's guidance, the following guiding principles were developed for assessing the biodegradability of flushable products.

- The product and its constituents should be inherently and practically biodegradable in household onsite treatment, municipal treatment, or relevant receiving environmental compartments.
- The physical form of the product or its constituents should not limit biodegradation.
- The product constituents should degrade predominantly to carbon dioxide, methane (anaerobic conditions) or other natural by-products, and/or be incorporated into bacterial biomass (or other natural carbon pools) at a rate that precludes accumulation of the material or its metabolites in receiving compartments. Determination of such a rate is best achieved by benchmarking against materials generally regarded and accepted as biodegradable (e.g., existing cellulosics).

When assessing the biodegradability of a product or its constituents, one needs to answer a series of questions. The questions for assessing biodegradability are presented in Figure 2-5.

The testing strategy for assessing biodegradability, which is based on the questions in Figure 2-5, is presented below. This strategy is used to generate product-specific evidence for support of biodegradability claims. For each question, the approach starts with rapid screening tests and, if needed, proceeds to more realistic laboratory tests. This strategy allows for the product developer to conduct the most appropriate amount of testing in support of a biodegradability claim.

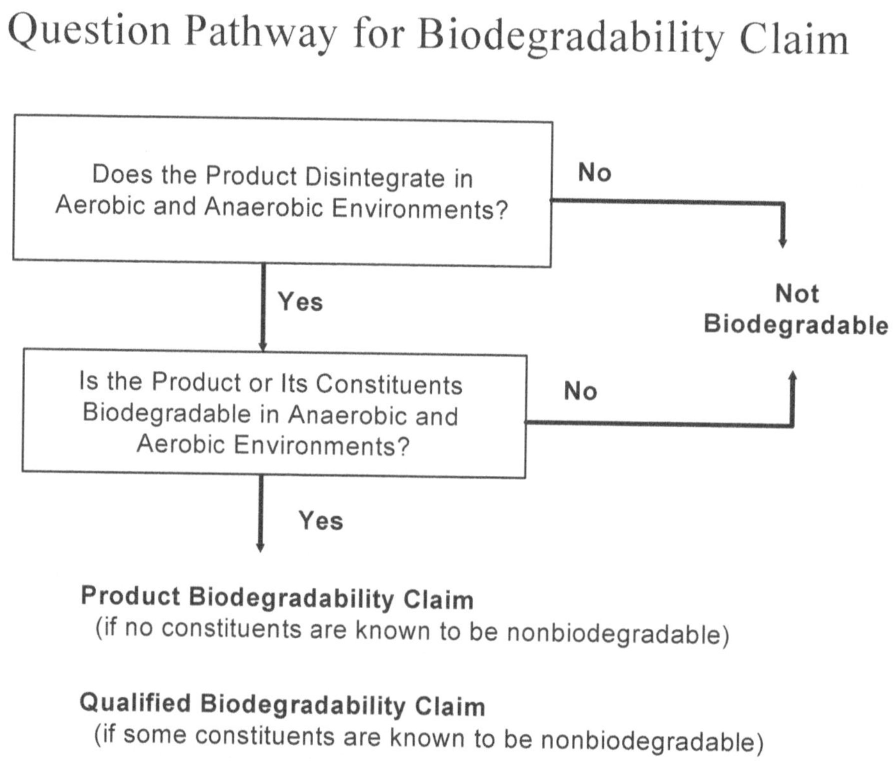

Figure 2-5. Question flow chart for assessing the biodegradability of a product or its constituents

The first step is to assess whether a product disintegrates in both aerobic and anaerobic environments. While the disintegration tests do not establish biodegradability *per se* because gas production is not measured, they do provide an indication of the potential for a product to biologically degrade. Since the disintegration tests would have most likely been performed for a flushability assessment, they can thus be used as a screening tool to determine whether a product should undergo biodegradation testing. The testing strategy to answer the flow chart questions in Figure 2-5 on biodegradability is presented below.

Disintegration:

The question below addresses the likelihood that a product or its constituents are biodegradable (Figure 2-5).

Question: Will the product or its constituents disintegrate in aerobic and anaerobic environments?

Testing Strategy
Tier 1–Anaerobic Disintegration Test
Tier 1–Activated Sludge Disintegration Test (Proposed)
Tier 2–Porous Pot Disintegration Test
Tier 3–Municipal or Onsite Mesh Bag Degradation Test

Biodegradation:

The question below addresses a product's ability to pass a readily biodegradability test (Figure 2-5). If the product does not pass these tests, higher tier testing could be performed to support biodegradability.

Question: Will the product or its constituents biodegrade in aerobic and anaerobic environments?

Testing Strategy
Tier 1–Aerobic Biodegradation Test
Tier 1 –Anaerobic Biodegradation Test

2.4 Quality Assurance

As with any scientific, data-based decision, those discussions about assessing the fate of flushable consumer product are only as reliable as the integrity of the data in support of such statements. Using principles of Total Quality Management and Quality Assurance, in conjunction with sound scientific investigation, users of test methods stated herein will be able to make sound conclusions. In many situations, flushability of consumer products is monitored by local or national governments. Use of Quality Assurance principles provides a basis for a user to present objective evidence that one has met the intent of applicable regulations and guidelines.

For many products, quality has been defined as "fitness for use." One can consider a product fit for use if it consistently performs the function for which it was designed and fulfills the needs and expectations of the consumer, including those related to flushability and environmental impact. Quality, in this regard, occurs when the development and production processes consistently meet well-defined standards.

Quality must be built into the product, first through properly designing and testing products and processes, and then by faithfully executing these proven designs. The people responsible for developing, manufacturing, and handling products are ultimately responsible for that products' continuing quality. The people responsible for marketing and selling must accurately represent products to customers and consumer so each person receives exactly what is expected.

For assessments of flushability and biodegradability several principles of Quality Assurance must be met in order to rely on data produced from studies regarding this topic.

Training:

Scientists in the testing laboratory must be trained on the method being used. Evidence is needed whereby the user of the method demonstrates proficiency in accurately producing results.

Written Procedures:

The test methods to be used in the laboratory must be written down in order to facilitate further development of the method or to ensure accurate, reproducible data over time and/or across users.

Validation:

The practice of validation is necessary to ensure the precision and accuracy of the testing data is true. Validation activities include assurances that equipment has been installed and is operating as intended. Validation also includes checks that ensure testing by one scientist is substantially equivalent to that of another scientist, using the same test method.

Laboratory Controls:

Many aspects of the testing lab need to be controlled so that results can be supported and repeated, if necessary. Testing equipment must be calibrated so that the results obtained from that piece of equipment are known to be precise and accurate. Reagents used in the testing laboratory are to be of known quality and source. Reagents must be labeled in order to prevent misuse and confusion in the laboratory.

Records:

The saying, "If it is not written down, it did not happen," is true in this testing scheme as well. Records provide the objective evidence or proof that a particular product design meets the needs of the consumer.

Internal Assessment Program:

Periodic reviews of the data systems used in assessing the fate of flushable product are a must. Only through periodic reviews can one ensure that methods are being used and implemented as intended. Internal assessments should provide a basis for continuous improvement in the testing lab.

Reliable data can only stem from studies and labs that implement these principles of Quality Assurance.

CHAPTER 3.0

TIER 1A: RAPID SCREENING TESTS

The Tier 1A tests are designed to provide a rapid screening for assessing the acceptability of a flushable product. These tests assess the compatibility of consumer products with household plumbing, wastewater conveyance, and wastewater treatment. The purpose of each test as well as a brief description of the method, and the acceptance criteria is presented below. If any of these test criteria are not met, additional higher tier testing may be needed to support flushability. The test methods may be found in Appendix A.

3.1 Screening Toilet Bowl and Drainline Test

3.1.1 Purpose of Test

The purpose of this test is to assess whether a product will pass through a toilet and household drainage pipe system without causing significant clogging under high usage conditions.

3.1.2 Method Description

This test method uses a toilet and drainline system that meets minimum U. S. national plumbing code requirements. The toilet being used (Kohler K-3490-EB) was obtained in the mid-1990s when low-flush toilets (1.6 gal) were first introduced. The toilets manufactured during this time period historically had poor performance. Thus, this toilet is used to represent a worst-case condition. This toilet relies on the establishment of a siphon in the trap to draw material out of the bowl. The product must then pass through the trap, which has a diameter of 1-5/8″ (40 mm). Once the product passes the trap, it enters the simulated drainage pipe system. The drainline consists of a 4″ (100 mm) diameter pipe with a slope of 2% (1/4″ per foot). The drainline is > 60′ (18 m) in length with four 90° bends, which adds conservatism to the test. An evaluation typically looks at several trials of five flushes per trial. These trials may consist of product only, product with another product, and/or product with synthetic fecal matter (SFM). The product loading to the toilet is targeted for the 90th percentile user, based on habits and practices data. If some of the material fails to clear the bowl or trap, a second flush is made. If the material remains in the bowl or trap after the second flush, then that flush is considered a failure. The material is then removed from the toilet via plunging and the sequence is continued. If the water level rises to the rim of the bowl, then that flush is considered a failure.

3.1.3 Acceptance Criteria

The acceptance criteria for this test are that a flushable product must: 1) clear the bowl and trap on four out of the five flushes, 2) travel more than 17′ (5 m) in the drainline on four out of five flushes, and 3) continues to travel more than 34′ (10 m) in the drainline after two flushes on four out of five times.

3.2 Column Settling Test

3.2.1 Purpose of Test

The purpose of this test is to assess the settleability of a flushable product in engineered clarification systems. A high rate of settling provides evidence that the product will be retained in a septic tank, removed in a grit chamber, and/or removed in the primary settling tank.

3.2.2 Method Description

This test is designed to assess the settling behavior of flushable products. The test apparatus consists of a transparent column that is mounted vertically on a test stand. The column has an inside diameter of 8″ (20 cm) and a depth of 60″ (152 cm). The diameter was chosen to minimize any sidewall effects on the settling of the test material. The depth was chosen to represent a septic tank. This test system is typically dosed following the product being flushed through the toilet and drainline system. Five replicate samples are dosed to the settling column, thus accounting for variability. The endpoint of this test is the time it takes for the test material to settle 45″ (115 cm). This settling rate is then used to assess whether the product will be retained in a septic tank, grit chamber or primary settling tank.

3.2.2 Acceptance Criteria

The acceptance criterion for this test is that the product must settle at a rate > 0.1 cm/s. This minimum rate of settling was determined using a CFD model developed specifically for assessing product retention in a septic tank.

3.3 Aerobic Disintegration Shake Flask Test

3.3.1 Purpose of Test

The purpose of this test is to assess the disintegration of a flushable product in wastewater conveyance systems. A high rate of disintegration provides evidence that the flushable product will be unrecognizable from other wastewater solids as it enters a municipal wastewater treatment plant or is discharged in combined sewer overflow.

3.3.2 Method Description

The aerobic disintegration shake flask test is similar to previously developed die-away shake flask methods (e.g., ASTM). It differs from most die-away test methods in that the endpoint is product weight loss over time instead of chemical loss over time. Moreover, the loss mechanism for a product in this test is mostly due to physical forces as opposed to biodegradation in a chemical die-away test. In this test, 3-5 g of test material is incubated in a 2.8 L baffled Fernbach flask that contains 1 L of prescreened (1 mm sieve) raw wastewater. Replicate flasks are then placed on a rotary shaker table and swirled at 150 rpm. This motion is used to simulate the turbulent force during wastewater conveyance. Periodically (6 hours, 24 hours and 48 hours), a flask is removed and its content is passed through a series of sieves (8 mm, 4 mm, 2 mm and 1 mm). At an initial screening level, only one flask is sampled at each time point. For confirmatory testing, triplicate samples are analyzed at each time point. The material retained on each sieve is collected, dried, and weighed. These weights are compared to the original test sample weight and are used to determine the rate of disintegration for various size fractions over time. While not specifically used in the acceptance criteria, the various size fractions lend insight into the rate and extent of disintegration. Moreover, many state and federal agencies define "trash" in waterways as material retained on a 5 mm mesh screen (e.g., California TMDL). Having information on the various size fractions would allow one to compare these results to governmental trash definitions, as well as to assess removal by screening.

3.3.3 Acceptance Criteria

The acceptance criterion for this test is for > 95% of the product mass to pass through a 1 mm sieve after 48 hours.

3.4 Anaerobic Disintegration Test

3.4.1 Purpose of Test

The purpose of this test is to assess the biological disintegration of a flushable product under anaerobic conditions. A high rate of disintegration provides evidence that the product will not be recognizable in septic tank septage or anaerobic digester sludge.

3.4.2 Method Description

This test is used to assess the biological disintegration of a flushable product under anaerobic conditions. In this test, the product is weighed and added to a 2 L amber glass bottle that contains 1.5 L of anaerobic digester sludge. The bottles are capped with a one-hole stopper to allow for venting of the evolved gases. At the screening level, three replicate bottles are placed in an incubator at 35°C for each product in the test. For confirmatory testing triplicate bottles are prepared for each time point. The test is conducted under static conditions. Periodically (7 days, 14 days and 28 days) a bottle is sacrificed and the content passed through a 1 mm sieve. The material retained on the 1 mm sieve is dried and weighed, and the percent of product mass loss is determined.

3.4.3 Acceptance Criteria

The acceptance criterion for this test is for > 95% of the product mass to pass through the 1 mm sieve after 28 days.

3.5 Activated Sludge Disintegration Test (Proposed)

3.5.1 Purpose of Test

The purpose of this test is to assess the biological disintegration of a flushable product under aerobic conditions. A high rate of disintegration provides evidence that the flushable product will persist in aerobic environments.

3.5.2 Method Description

The activated sludge disintegration test is similar to previously developed die-away shake flask methods (e.g., ASTM). It differs from most die-away test methods in that the endpoint is product weight loss over time instead of chemical loss over time. In this test, 3–5 g of test material is incubated in a 2.8 L baffled Fernbach flask that contains 1 L of prescreened (1 mm sieve) activated sludge. Replicate flasks are then placed on a rotary shaker table and swirled at 150 rpm. This swirling motion is used to keep the test system aerobic. Periodically (7 days, 14 days and 28 days), a flask is removed and its content is passed through a 1 mm sieve. At an initial screening level, only one flask is sampled at each time point. For confirmatory testing, triplicate samples are analyzed at each time point. The material retained on the sieve is collected, dried, and weighed. These weights are compared to the original test sample weight and are used to determine the rate of disintegration over time.

3.5.3 Acceptance Criteria

Because this test is still under development, no acceptance criteria have been developed at this time.

CHAPTER 4.0

TIER 1B: MATERIAL PROPERTIES TESTS

The Tier 1B test methods are used to assess the rate and extent of product degradation via physical, chemical and biological processes. The outcome of these tests is dependent on the material properties of the product being tested. The test methods may be found in Appendix A.

4.1 Aerobic Biodegradation Test

4.1.1 Purpose of Test

The purpose of this test is to assess the aerobic biodegradability of a product or its constituents. A high rate and extent of biodegradation provides evidence that the product will not persist in the environment for an extended period of time.

4.1.2 Method Description

An electrolytic respirometer, or an equivalent apparatus, is used to measure the rate and extent at which a test material undergoes aerobic biodegradation. This test typically uses comminuted material, but whole test product could also be tested. The respirometer continuously and automatically measures the amount of O_2 taken up by microbes, as well as the amount of CO_2 produced. The test material, media, and inoculum (activated sludge) are incubated at ± 22°C in a flask connected to a sensor head that contains a CO_2 probe, a Cu_2SO_4 reservoir, and a pressure transducer. As CO_2 is liberated from the media, it is absorbed by the alkali trap of the CO_2 probe. The change in conductivity of the base solution is automatically recorded. As the O_2 is consumed by the microbes, the pressure in the head space is reduced. This reduction in pressure is detected by the pressure transducer, which triggers the electrolysis of the Cu_2SO_4 and in turn liberates free O_2. The amount of O_2 needed to restore this pressure is proportional to the magnitude and duration of the electrolyzing current. The extent of biodegradation is evaluated by determining the amount of CO_2 that evolved from the product, which is determined from the amount of carbon in the test material.

4.1.3 Acceptance Criteria

The acceptance criteria for this test method is that there needs to be > 60% CO_2 evolution over a 10-day window (after 10% CO_2 evolution is achieved) within the 28-day test period. These criteria come from the OECD 301B test. If these criteria are met, then the test material is considered readily biodegradable. If < 10% CO_2 is measured, the test material is considered nonbiodegradable. A CO_2 value between 10% and 60% suggests that some biodegradation is occurring, but there may either bc a slowly degradable intermediate being formed or some of the constituents in the test material are not biodegradable. Additional testing may be warranted with this result.

4.2 Anaerobic Biodegradation Test

4.2.1 Purpose of Test

The purpose of this test is to assess the anaerobic biodegradability of a product or its constituents. A high rate and extent of biodegradation over the 56-day test period provides evidence that a flushable product will biodegrade in septic tanks and anaerobic sludge digesters.

4.2.2 Method Description

This test method is similar to other anaerobic biodegradability tests developed for down the drain chemicals (e.g., ASTM). The major difference of this test method is that a flushable product or its constituents are used in the test instead of a specific chemical. This test typically uses comminuted material, but whole test product can also be tested. In this test, four replicate 160 mL serum bottles are filled with 50 mL of test media and inoculum (anaerobic digester sludge). The test material is added at a concentration of 50 to 300 mg/L. Preparation of the media and inoculum are performed under a nitrogen blanket to ensure anoxic conditions. Gas production is continuously measured for each serum bottle with a pressure transducer. A blank without product is also included so that background respiration can be measured. The serum bottles are incubated at 35°C and anaerobic conditions are monitored throughout the test with resazurin. At the end of the test, the dissolved inorganic carbon concentration in the media is determined and the amount of dissolved CO_2 is corrected in the gas phase. The theoretical anaerobic biodegradation at 56 days is reported, which is based on a measured percent carbon in the product prior to test initiation.

4.2.3 Acceptance Criteria

The acceptance criterion for this test is > 70% gas evolution over a 56-day period. A test material that passes this criterion would be considered biodegradable. If there is < 10% gas evolution, then the material is assumed to be nonbiodegradable under anaerobic conditions. Gas production between 10% and 70% indicates that some biodegradation has occurred, but there may either be a slowly degradable intermediate being formed or some of the constituents in the test material are not biodegradable.

4.3 Physical Disintegration Test

4.3.1 Purpose of Test

The purpose of this test is to assess the dispersability of a flushable product in household and municipal conveyance systems. A high rate of dispersability would indicate that a product would readily break up in conveyance systems.

4.3.2 Method Description

This test is used to gain a better understanding of the physical disintegration of a product. In this test a standard 2 L beaker is filled to the 2 L mark with tap water and placed on a magnetic stir plate. The product is then placed in the beaker of water and after five seconds the magnetic stirrer is turned on. The speed of the stir bar is set at 650 rpm. When the first piece of product breaks off, the timer is stopped and the time to initial break up is recorded. Five replicates are run for each test material and the average time to break up is reported. If the test material does not begin to break up after 10 minutes, the test is terminated.

4.3.3 Acceptance Criteria

There are no acceptance criteria for this test because it is used to provide insight into the loss mechanism.

4.4 Hydrolysis/Solubility Test

4.4.1 Purpose of Test

The purpose of this test is to assess the rate at which a product hydrolyzes or solubilizes in an aqueous solution. Results from this test will help assess the loss mechanism of a flushable product.

4.4.2 Method Description

This test method is similar to the aerobic disintegration shake flask test except that the flasks are not swirled and tap water is used instead of raw wastewater. In this test, 3-5 g of test material is incubated in a 2.8 L baffled Fernbach flask that contains 1 L of tap water. Periodically (6 hours, 24 hours and 48 hours), a flask is removed and its content is passed through a series of sieves (8 mm, 4 mm, 2 mm and 1 mm). The material retained on each sieve is collected, dried, and weighed. These weights are compared to the original test sample weight, which provides for an evaluation of the rate of disintegration for various size fractions.

4.4.3 Acceptance Criteria

There are no acceptance criteria for this test because it is used to provide insight into the loss mechanism.

CHAPTER 5.0

TIER 2: CONFIRMATORY TESTS

The Tier 2 confirmatory tests are designed to be more realistic than the Tier 1 tests for assessing the fate and compatibility of consumer products in wastewater disposal systems. If any of the acceptance criteria in these tests are not met, then additional higher tier testing may be needed to support flushability. The test methods may be found in Appendix A.

5.1 Confirmatory Toilet Bowl and Drainline Test

5.1.1 Purpose of Test

The purpose of this test is to assess whether a flushable produce will pass through a toilet and household drainline system without causing significantly clogging under normal usage conditions.

5.1.2 Method Description

This test method uses the same toilet and drainline system as described in the Tier 1A rapid screening test (see description in Section 3.1). A confirmatory evaluation typically simulates two days of normal toilet use by a family of four (based on habits and practices data). The flushing condition includes male urination (no product), female urination (with bath tissue), bowel movement (bath tissue along with simulated fecal material), and any other product of interest. A total of 34 flushes are made per test run. Triplicate test runs are performed for each product type. If some of the material fails to clear the bowl or trap, a second flush is made. If the material remains in the bowl or trap after the second flush, then this is considered a failure. The material is then removed from the toilet by plunging and the sequence is continued. If the water level rises to the rim of the bowl, then that flush is considered a failure.

5.1.3 Acceptance Criteria

The product is considered acceptable if > 95% of the flushes are successful and the product continues to move through the drainline on each subsequent flush. In addition to these criteria, results from a particular run are compared to data from acceptable products that are currently being marketed.

5.2 Tipping Pipe Disintegration Test (Proposed)

5.2.1 Purpose of Test

The purpose of this test is to provide for a more realistic simulation of the fate of a flushable product during wastewater conveyance. This test also provides insight in support of the Tier 1A Shake Flask test. A high rate of break up in this test indicates that the product will be indistinguishable from other wastewater solids entering a wastewater treatment plant.

5.2.2 Method Description

This method uses flowing water in a pipe to assess the physical disintegration of a product. In this test, a flushable product is placed in a 6″ (15 cm) diameter, 20′ long (6 m) clear plastic pipe. The water in the pipe is rocked back and forth by changing the pipe slope +/- 3% over a 30-second period. Numerical model simulations were used to determine the slope that would best represent the velocities and turbulent forces in a wastewater conveyance system. This slope may change as additional data becomes available for turbulent forces in real-world sewer pipes. At specified sampling times the content in the pipe is passed through a series of sieves (8 mm, 4 mm, 2 mm, and 1 mm) and the material retained on each sieve is collected, dried, and weighed. Weights from each size fraction are then compared to the original test material weight and an evaluation of disintegration over time is made. While not specifically used in the acceptance criteria, the various size fractions lend insight into the rate and extent of disintegration. Moreover, many state and federal agencies define "trash" in waterways as material retained on a 5 mm mesh screen (e.g., California TMDL). Having information on the various size fractions would allow one to compare these results to governmental trash definitions, as well as to assess removal by screening.

5.2.3 Acceptance Criteria

Because this test is still under development and its ultimate use may be to bridge the Tier 1 Shake Flask test with the Tier 3 Field Spiking test, no acceptance criteria have been developed at this time.

5.3 Porous Pot Disintegration Test

5.3.1 Purpose of Test

The purpose of this test is to assess the fate of a flushable product during aerobic onsite treatment and municipal activated sludge treatment. A high rate of disintegration would indicate that biological degradation of the product is occurring.

5.3.2 Method Description

This test method is similar to other continuous flow activated sludge tests that have previously been developed for down the drain chemicals. However, this test differs from those tests in that the endpoint is loss of product mass in the system versus loss of a specific chemical through the treatment process. The test apparatus consists of a 6 L glass reactor with a porous stainless steel filter. The stainless steel filter is used to retain the activated sludge solids in the reactor. Raw wastewater is continuously fed to the reactor at a rate of approximately 15 mL/min. This corresponds to a hydraulic retention time (HRT) of about seven hours. The mixed liquor suspended solids (MLSS) are periodically measured and a portion of the solids wasted on a weekly basis to maintain the MLSS between 2500 to 4500 mg/L. In this test, preweighed product is placed in mess bags (fiberglass screening with a 1.6 mm size opening) and then suspended in the porous pot reactor. At designated time points, one of the mesh bags with product is removed from the reactor and its content rinsed through a 1 mm sieve. The amount of material remaining on the 1 mm sieve is then dried and weighed. The loss of product mass is determined over time.

5.3.3 Acceptance Criteria

A weight loss of > 95% over a 28-day period (approximately three SRT in an activated sludge treatment plant) would indicate that the product is most likely biodegradable and compatible with aerobic biological treatment.

5.4 Laboratory Soil Disintegration Test (Proposed)

5.4.1 Purpose of Test

The purpose of this test is to assess the disintegration of a product in soil following septage or sludge (biosolids) amendment. A rapid loss rate indicates that a product will not be recognizable in a septage or sludge amended soil scenario.

5.4.2 Method Description

In this test, the product is placed in mesh bags (fiberglass screening with a 1.6 mm size opening) and then buried in soil or amended directly to the soil surface. These placements of the mesh bags would simulate subsurface incorporation and surface application. In either scenario, the product can also be amended to the soil with septage or sludge. The test is conducted in a soil box in the laboratory. The dimension of the experimental soil box is 2′ × 2′ (0.6 × 0.6 m) and the depth of the soil is about 6″ (15 cm). The soil in the box is comprised of a mixture of topsoil and peat moss. During the test, 1″ (2.5 cm) of tap water is added weekly to each soil box to simulate rainfall. Periodically, a product mesh bag is retrieved from the soil and its content rinsed through a 1 mm sieve. The material remaining on the 1 mm sieved is dried and weighed. The percent of product weight loss over time is calculated.

5.4.3 Acceptance Criteria

While this test method is still under evaluation, a weight loss of > 95% over a two month period would indicate that the product would not be recognizable in the soil environment over an extended period of time.

5.5 Sewage Ejector and Grinder Pump Test

5.5.1 Purpose of Test

The purpose of this test is to evaluate the compatibility of flushable products with household wastewater ejector and grinder pumps. This test is only used to assess the flushability of bath tissue because most pump manufacturers make statements in their warranties to not flush rags, towels, cloths, tampons or feminine protection products.

5.5.2 Method Description

Residential plumbing systems typically rely on gravity for the conveyance of wastewater. There are some instances, though, in which pumps are needed to either lift wastewater so that it can be transported by gravity (ejector pumps) or be pumped into a pressurized sewer system (grinder pumps). In this test, a flushable product is dosed in tap water to a catch basin where either the ejector or grinder pump resides. The pump is periodically activated, depending on the water-level sensor, and a portion of the basin's content is evacuated. The evacuated content is then pumped to a storage basin that is located approximately 8′ (2.4 m) above the pump basin. This elevated basin provides a pressure head for the pumps to lift against. The content in the upper basin is then screened prior to being reused in the lower basin. The system is loaded at a rate equivalent to one week of product use (based on habits and practices data) or until the system fails, whichever comes first.

5.5.3 Acceptance Criteria

A flushable product is deemed acceptable if it can pass through the pump without clogging at a load to the basin equivalent to one week of product use. Failure occurs when the pump no longer functions properly with the water rising to the top of the basin.

CHAPTER 6.0

TIER 3: FIELD VALIDATION TESTS

This chapter presents the field validation test methods. These tests are generally designed to assess the compatibility of flushable products in wastewater disposal systems under realistic conditions. This is the ultimate tier for assessing the flushability of consumer products. Thus, failure to meet the acceptance criteria may mean that a qualifier claim like "Not Recommend for Septic Systems" or a warning statement like "Do Not Flush" will be made for the product. The test methods are presented in Appendix A.

6.1 Home Use Toilet and Drainline Test

6.1.1 Purpose of Test

The purpose of this test is to assess the compatibility of a flushable product with household plumbing fixtures under normal usage conditions.

6.1.2 Method Description

In this test, the flushable product is provided to several households (typically 50-100) and the household members are asked to use the product as they normally would. These households are selected based upon their responses to a pretest screening questionnaire. Criteria for participation typically includes family size (three or more in household), type of toilet (age), type of drainage pipe (plastic, cast iron, etc.), type of treatment system (onsite or municipal), and data concerning product use. The in-life portion of the test typically lasts from one to two months, but it could run longer depending on the product being tested. Household members are asked to keep a diary during the test period, which includes how much product is being used and if any problems occurred. At the end of the study period the household is asked to fill out a post-test questionnaire. The results are tabulated in terms of the number of incidents per person per month or by the number of incidents per product use per month.

6.1.3 Acceptance Criteria

If the product type being tested is already on the market (e.g., bath tissue), then the results of the new product are compared or benchmarked against the currently marketed product. If the number of incidents is comparable to that of the currently marketed product, then the new product is deemed acceptable. If the product type is new to the world, then its acceptability is determined by comparing the number of incidents to a similar product already on the market.

6.2 Onsite Treatment—Septic System Test

6.2.1 Purpose of Test

The purpose of this test is to assess the fate of flushable products in a home onsite septic system.

6.2.2 Method Description

In this test, the flushable product is evaluated in a full-scale household septic tank. While the description below presents the test method for one site, the method could be easily expanded to multiple sites if needed. The septic tank has been instrumented with two access ports (one at the inlet and the other near the outlet of the tank), as well as an effluent trap on the outlet drainage pipe. Two types of fate testing are conducted at the field site. These include a mesh bag disintegration test and a product retention test. In the disintegration test, the product is placed in a fiberglass mesh bag and tethered down into the tank. Periodically, a mesh bag is removed and its content rinsed through a 1-mm sieve. The material retained on the 1-mm sieve is then dried and weighed. The weight loss of the product over time is then determined. For the product retention test, the product is flushed in the home over a specified period of time (e.g., one week). The retention of the product in the tank is then monitored over at least a 6-month period by making observations in the outlet effluent trap and the tank observation ports. In addition, on approximately a yearly basis or after the testing is completed, the content of the tank is pumped and passed through a basket screen (approximately 1″ opening) and catch screen (approximately 1/8″ opening) to assess how much of the product remains in the tank. The basket screen simulates the type of screening a hauler might use when amending septage to soil and the catch screen collects the products that visually end up in septage-amended soil. If any product is identified in the septage that passes through the basket screen, then additional soil disintegration testing is needed (Section 5.4). The soil disintegration test is used to assess the product's fate in a septage amended soil scenario. It is assumed that any product retained on the basket screen would be landfilled.

6.2.3 Acceptance Criteria

The product is acceptable for septic systems if none of the product escapes the tank over a six-month period following dosing and if none of the product is visible in the septage following pumping and screening prior to septage land application. If visible product passes through the basket screen, then the product must disintegrate over a reasonable period of time in a soil disintegration test to be acceptable.

6.3 Onsite Treatment—Aerobic System Test

6.3.1 Purpose of Test

The purpose of this test is to assess the compatibility of a flushable product in a home onsite aerobic treatment system.

6.3.2 Method Description

In this test the flushable product is evaluated in a full-scale household aerobic treatment system. While this method describes the approach for one system, it could easily be expanded to multiple sites if warranted. The onsite aerobic system used in this test has been instrumented with three access ports (one in the inlet settling tank, one in the aeration unit, and one in the outlet of the system). An effluent trap on the outlet drainage pipe from the aeration unit, which is prior to sand filtration, was also installed. Two types of fate testing are conducted at the field site, i.e., a mesh bag disintegration test and a product compatibility test. In the disintegration test, a product is placed in fiberglass mesh bags and tethered down into the aeration chamber. Periodically, a mesh bag is removed and its content rinsed through a 1-mm sieve. The material retained on the 1-mm sieve is then dried and weighed. The weight loss of the product over time is then determined. In the treatment process compatibility test, the product is flushed in the home and its fate is monitored in the outlet effluent trap and the tank observation ports over at least a 6-month period. In addition, on approximately a yearly basis or after the testing is completed, the content of the system (settling and aeration tanks) is pumped and then passed through a basket screen (approximately 1″ opening) followed by a catch screen (approximately 1/8″ opening). The basket screen simulates the screening a hauler might perform prior to applying the septage to land, whereas the catch screen retains the visible material that would be amended to land. If any product is identified in the septage that passes through the basket screen, then soil disintegration testing is needed (Section 5.4) to assess the product's fate in a septage amended soil scenario. It is assumed that any product retained on the basket screen would be landfilled.

6.3.3 Acceptance Criteria

The product is acceptable for onsite aerobic systems if none of the product escapes the aeration tank over a six-month period and if none of the product is identifiable in the septage after it has passed through the basket screen prior to being land-applied. If visible product passes through the screen, then the product must disintegrate over a reasonable period of time in a soil disintegration test to be acceptable.

6.4 Sewer Conveyance/Municipal Treatment Test

6.4.1 Purpose of Test

The purpose of this test is to determine the fate and transport of a flushable product in a municipal wastewater conveyance and treatment system.

6.4.2 Method Description

In this test, a flushable product is either spiked into the headworks of a municipal wastewater treatment plant or spiked into the sewer collection system at a distance of about six hours travel-time away from the headworks of the municipal wastewater treatment plant. For the headworks spiking test, the fate and transport of the product is monitored throughout the treatment plant over a two-day period. Hourly observations of the unit processes are typically made over the first 12 hours and at six-hour intervals thereafter to record the time, location, and condition of the product. Several locations throughout the treatment plant are monitored (e.g., headworks, bar racks and scraping belts, grit chamber, primary settling tank, scum pits, aeration units, secondary settling tanks, wasted sludge). The bar rack scraping belt is typically turned off between observations to ensure that the product being screened is counted. For the sewer-spiking test, the product is monitored over a three-day period throughout the treatment plant. In both of these tests, anywhere from 500 to 2000 products are spiked into the system to ensure that a sufficient number can be tracked through the collection and treatment system. It is assumed that product removed by the bar rack, grit chamber or scum removal systems become solid waste and are disposed at a landfill. Product that settles in the primary settling tank will enter the sludge digester and thus could become a part of the biosolids being land-applied. Product disintegration can be assessed via mesh bags in the aerobic and anaerobic digesters. Moreover, if product has been dosed over a longer period of time (e.g., during a test market study), then the biosolids and the biosolids-amended soil can be visually inspected for any evidence of the product.

6.4.3 Acceptance Criteria

To be acceptable, a product should be effectively removed by the bar rack, grit chamber or during primary treatment, and should not cause a nuisance to the operation of the treatment plant. If the product is observed in the effluent of the treatment plant at a level > 0.5% of the spiked amount, then that product would need to pass the Tier 1B Aerobic Biodegradation test to be considered compatible with municipal wastewater treatment. If the product is observed in the digested sludge (biosolids) of the treatment plant at a level > 0.5% of the spiked amount, then that product would need to pass the Tier 2 Soil Disintegration test to be considered compatible with municipal wastewater treatment.

6.5 Field Soil Disintegration Test

6.5.1 Purpose of Test

The purpose of this test is to assess the disintegration of a product in soil following septage or sludge (biosolids) amendment. A rapid loss rate indicates that a product will not be recognizable in a septage or sludge amended soil scenario.

6.5.2 Method Description

In this test, the product is placed in mesh bags (fiberglass screening with a 1.6 mm opening) and then buried in soil or amended directly to the soil surface. These placements of the mesh bags would simulate subsurface incorporation and surface application. In either scenario, the product can also be amended to the soil with septage or sludge. The dimension of the experimental soil plot is 6′ x 8′ (1.8 x 2.4 m) and the depth of the soil is about 6″ (15 cm). The soil in the plot is comprised of a mixture of topsoil and peat moss. Weekly precipitation is measured with a rain gauge and if < 1″ (2.5 cm) of rain is observed, then a sufficient amount of water is added to the plot to ensure that at least 1″ (2.5 cm) of rain is dosed during the summer months. Periodically, a product mesh bag is retrieved from the soil and its content rinsed through a 1 mm sieve. The material remaining on the 1 mm sieved is dried and weighed. The percent of product weight loss over time is calculated.

6.5.3 Acceptance Criteria

While this test method is still under evaluation, a weight loss of > 95% over a two-month period (summer months) would indicate that this product would not be recognizable in the soil environment over an extended period of time.

CHAPTER 7.0

Approach Overview and Case Studies

7.1 Overview of Approach

The flushability of a consumer product, which includes its fate and compatibility with wastewater disposal systems, should be assessed prior to being acceptable for market. In the approach presented in this document, a flushable consumer product is evaluated by a tiered testing approach that consists of a series of increasingly complex laboratory and field tests as needed. This tiered evaluation could ultimately include testing the product in full-scale field systems. While evaluating a consumer product in a full-scale system is the most realistic method, it is not always feasible. Thus, laboratory tests are usually performed first and their results extrapolated to full-scale systems.

The ability of a product to pass through household toilets and drainline systems is evaluated first. If the product has difficulty passing through a toilet and drainline, then the testing is terminated and the product is considered unacceptable as a flushable product. If the product passes through a household plumbing system, the next step is to evaluate its compatibility with municipal and onsite wastewater treatment systems. Key factors to consider in this evaluation are the settleability, physical strength, and biodegradability of the product. Settleability can be evaluated by knowing the product density or by performing a column-settling test. Product strength can be evaluated by wet tensile strength measurements or by physical disintegration testing (e.g., aerobic shake flask test). Biological disintegration of a product can be evaluated by measuring weight loss after exposure to microbes. This type of testing does not establish biodegradability *per se* because gas production is not measured, but it does provided an indication of the potential for biological degradability. Aerobic biodegradation can be determined via oxygen uptake or carbon dioxide production. An electrolytic respirometer is used for this purpose. Respirometry provides information on the rate and extent of biodegradation, possible inhibition to biological activity, and oxygen demand of the test material. A serum bottle test is used to assess gas production (methane and carbon dioxide) under anaerobic conditions, thus providing evidence of anaerobic biodegradation. Field tests can be used to confirm the laboratory results.

7.2 Testing Strategy

Since a product may undergo different disposal pathways depending on the wastewater treatment infrastructure available to the household (e.g., municipal vs. onsite treatment), a different testing strategy is needed for each of the disposal pathways. These various disposal pathways were previously depicted in Figure 2-1 and the questions to be answered in Figures 2-2 through 2-4. The testing strategy starts with the home and then proceeds to the various wastewater treatment systems. In this strategic approach all of the testing data, along with weight of evidence of currently marketed products, is used to evaluate the flushability of a product.

At the rapid screening level (Tier 1), five tests are used to assess the fate and compatibility of consumer products following the disposal through a toilet. These tests include: 1) a toilet and drainline test; 2) a column-settling test; 3) an aerobic shake flask disintegration test; 4) an anaerobic disintegration test; and 5) an activated sludge disintegration test (proposed) or aerobic biodegradation test. If the acceptance criteria are met for all five of these tests, the product is considered acceptable for market as a flushable product and no additional testing is needed. The screening toilet and drainline test would satisfy the household assessment in Figure 2-2, the shake flask test would fulfill the testing needed for the municipal wastewater collection and treatment assessment in Figure 2-3, as well as the ejector and grinder pumps in Figure 2-2. The column settling and anaerobic disintegration tests would meet the criteria for septic systems in Figure 2-4, and the column settling and activated sludge or aerobic biodegradation tests would meet the criteria for aerobic onsite treatment systems in Figure 2-5. Thus, these five screening tests answer all of the flushability questions presented in Figures 2-2 through 2-4 and they ensure that a product will not clog toilets or household drainlines under heavy use (95th percentile user), disintegrate rapidly during wastewater conveyance, settle in septic tanks and aerobic onsite systems, disintegrate rapidly under anaerobic conditions, or disintegrate rapidly under aerobic conditions. If any of the criteria are not met in these five tests, then additional higher tier testing is used to assess the product under more realistic conditions.

7.3 Case Studies for Flushable Products

Below are three examples of flushable products. These examples illustrate how the overall approach and test methods are used to assess flushability.

7.3.1 Bath Tissue

A new bath tissue is being developed for market. The bath tissue is comprised mostly of pulp fibers, but it does contain a temporary wet strength agent at a 2% (w/w) level. The new product will have a higher basis weight and wet strength than the currently marketed product. Product development would like to know if the new product is an acceptable flushable product.

7.3.1.1 Tier 1 Testing Results

Toilet and Drainline Test—Pass

The product was able to pass through the toilet and trap on 5 out of 5 flushes under a high loading (note acceptance criterion four out of five flushes), and was able to move > 5 m in the drainline after the first flush on five out of five flushes and > 10 m after the second flush (note acceptance criterion is criterion four out of five flushes).

Aerobic Shake Flask Test—Pass
The product physically disintegrated rapidly with > 95% of its mass passing through a 1 mm sieve after one hour (note acceptance criterion > 95% after 48 hr).

Column Settling Test—Pass
The product settled at a rate of 2 cm/s (note acceptance criterion > 0.1 cm/s).

Anaerobic Disintegration Test—Pass
The product disintegrated rapidly with > 95% of its mass passing through a 1 mm sieve in 7 days (note acceptance criterion > 95% in 28 days).

Aerobic Biodegradation Test—Pass
The product biodegraded rapidly under aerobic conditions with > 70% CO_2 evolution within a 10-day period (note acceptance criterion > 60% within 10-day window).

Anaerobic Biodegradation Test—Pass
The product biodegraded rapidly under anaerobic conditions with > 80% gas production within a 14-day period (note acceptance criterion > 70% gas production in 56 d).

7.3.1.2 Analysis

Based on these test results and its composition, the product would be considered flushable. No additional higher tier testing is needed. Because the product contains a nonbiodegradable wet strength agent at a 2% level, it would not be labeled biodegradable even though it passed both the aerobic and anaerobic biodegradation tests. Though, claims that the pulp fibers in this product are biodegradable could be made. Moreover, the product is expected to have minimal clogging in toilets and household drainlines and to be compatible with wastewater treatment systems (both onsite and municipal systems). For onsite treatment (both septic and aerobic systems) the product is expected to settle and disintegrate over a reasonable period of time. During wastewater conveyance the product is expect to disintegrate into small particles that would be indistinguishable from other wastewater solids. Once entering the municipal wastewater treatment plant the product would follow other wastewater solids. Some of the product would be removed in the grit chamber where it would subsequently be landfilled, some of the product would settle in the primary clarifier where it would enter the sludge handling process, and some of the product may enter the secondary biological treatment system where it would biodegrade. Once entering the sludge handling process the product is expected to biodegrade during either aerobic or anaerobic digestion. Since this product degrades rapidly in both aerobic and anaerobic environments, it is not expected to accumulate in wastewater treatment systems or cause a significant increase in maintenance requirements.

7.3.2 Wet Wipes

A new flushable and biodegradable wet wipe is being developed for market. This new product is comprised of 100% hydroentangled rayon (modified cellulose fibers) and will have a higher wet strength than the currently marketed pulp/binder wet wipe product. Product development would like to know if this new product is an acceptable flushable product. Moreover, they would like to know if biodegradable claims could be made.

7.3.2.1 Tier 1 Testing Results

Toilet and Drainline Test—Pass

The product was able to pass through the toilet and trap on five out of five flushes when disposed of as directed (note acceptance criterion is four out of five flushes), and was able to move > 10 m in the drainline on five out of five flushes (note acceptance criterion > 10 m after two flushes four out of five times).

Aerobic Shake Flask Test—Fail

The product did not undergo rapid physical disintegration with < 5% of its mass passing through a 1 mm sieve after 48 hour (note acceptance criterion > 95% after 48 hr).

Column Settling Test—Pass

The product settled at a rate of 3 cm/s (note acceptance criterion > 0.1 cm/s).

Anaerobic Disintegration Test—Pass

The product disintegrated rapidly with > 95% of its mass passing through a 1 mm sieve in 7 days (note acceptance criterion > 95% in 28 days).

Aerobic Biodegradation Test—Pass

The product biodegraded rapidly under aerobic conditions with > 60% CO_2 evolution within a 10-day period (note acceptance criterion > 60% within 10-day window). Moreover, there was > 80% CO_2 evolution after 21 days and 88% CO_2 evolution after 30 days.

Anaerobic Biodegradation Test—Pass

The product biodegraded rapidly under anaerobic conditions with 76% gas production within 32 days, which is when the test was terminated (note acceptance criterion > 70% gas production in 56 d).

7.3.2.2 Analysis

Based on these testing results, the product would be considered flushable and biodegradable. The biodegradability claim is supported by its rapid and extensive biodegradation under both aerobic and anaerobic conditions. Moreover, the product is expected to have minimal clogging in toilets and household drainlines when disposed of as directed, as well as to be compatible with onsite and municipal wastewater treatment systems. However, this product may contribute to clogging in household drainlines if an obstruction is present like a tree root. Keeping the household drainlines free of obstructions would be the homeowner's responsibility. Moreover, this product is not expected to be compatible with household ejector and grinder pumps. This warning is usually addressed in the pump manufacturer's warranty where statements are made such as "cloths and wash towels should not be disposed into these systems." For onsite treatment (both septic and aerobic systems), the product is expected to settle and disintegrate over a reasonable period of time. The wipes are not expected to disintegrate significantly during conveyance, but once they enter a municipal wastewater treatment plant the product is expected to be screened by the bar rack or grinded in the comminutor, followed by settling in the grit chamber where the material would eventually be landfilled. Any product that passes the grit chamber is expected to settle in the primary clarifier where it would enter the sludge handling process. The product would not be expected to enter the secondary biological treatment system, but if it did, it would be expected to biodegrade. Once entering the sludge handling process, the product is expected to biodegrade during either aerobic or anaerobic digestion. Since this product degrades rapidly in both aerobic and anaerobic environments, it is not expected to accumulate in wastewater treatment systems or cause a significant increase in maintenance requirements.

7.3.3 Tampons

A tampon is currently being marketed as flushable and biodegradable. This tampon is comprised of a cotton core, a rayon (modified cellulose fibers) overwrap, and cotton cord. The FTC has requested to see all available data that support the claim that this product is flushable and biodegradable.

7.3.3.1 Tier 1 Testing Results

Toilet and Drainline Test—Pass

The product was able to pass through the toilet and trap on five out of five flushes when disposed of as directed (note acceptance criterion is four out of five flushes), and was able to move > 10 m in the drainline on five out of five flushes (note acceptance criterion is > 10 m after two flushes on four out of five times).

Aerobic Shake Flask Test—Fail

The product did not undergo rapid physical disintegration with < 5% of its mass passing through a 1 mm sieve after 48 hour (note acceptance criterion > 95% after 48 hr).

Column Settling Test—Pass

The product settled at a rate of 2 cm/s (note acceptance criterion > 0.1 cm/s).

Anaerobic Disintegration Test—Pass

The product disintegrated rapidly under anaerobic conditions with > 95% of its mass passing through a 1 mm sieve after 28 days (note acceptance criterion > 95% in 28 days).

Aerobic Biodegradation Test—Conditional Pass

In an electrolytic respirometry test the amount of oxygen consumed to oxidize the product in comparison to its chemical oxygen demand (COD) was 99% after 53 days. While this test was performed prior to establishing the current test method, these results indicate that the product has a substantial potential for aerobic biodegradation.

Anaerobic Biodegradation Test—Pass

The cotton core and the rayon overwrap had > 70% gas production after 15 days, and the cotton cord had > 70% gas production after 60 days (note acceptance criterion > 70% in 56 days). The total gas production after 76 days for the cotton core, rayon overwrap and cotton cord were 90%, 83% and 75%, respectively.

7.3.3.2 Tier 3 Testing Results

Septic System Treatment—Confirmatory

A mesh bag study using several septic tanks demonstrated variable rates of degradation depending on temperature (i.e. season of the year). During the summer months the product showed complete degradation in a septic tank within 84 days.

Municipal Wastewater Treatment—Confirmatory

A mesh bag study in an anaerobic digester demonstrated a product weight loss of 97% within 21 days.

7.3.3.3 Analysis

Based on these testing results, the product would be considered flushable and biodegradable. The product is expected to have minimal clogging in toilets and household drainlines when disposed of as directed, as well as to be compatible with onsite and municipal wastewater treatment systems. However, this product may contribute to clogging in household drainlines if an obstruction such as a tree root is present. Keeping the household drainlines free of obstructions would be the homeowner's responsibility. Moreover, this product is not expected to be compatible with household ejector or grinder pumps. This warning is usually addressed in the pump manufacturer's warranty where statements are made such as "tampons and feminine protection products should not be disposed into these systems." For onsite treatment (both septic and aerobic systems), the product is expected to settle and disintegrate over a reasonable period of time. Once entering the municipal wastewater treatment plant, the product is expected to be screened by the bar rack or grinded in the comminutor, followed by settling in the grit chamber. This material would eventually be landfilled. Any product that passes the grit chamber should settle in the primary clarifier where it would enter the sludge handling process. This product would not be expected to enter the secondary biological treatment system, but if it did, it would be expected to biodegrade. Once entering the sludge handling process, the product is expected to biodegrade during aerobic or anaerobic digestion. Since this product degrades rapidly in both aerobic and anaerobic environments, it is not expected to accumulate in wastewater treatment systems or cause a significant increase in maintenance requirements.

Furthermore, this tampon has been marketed since 1936 with no known environmental damage. This weight of evidence is supported by the fact that the product is comprised of cellulosic fibers that are readily attacked and degraded by microbes found naturally in both aerobic and anaerobic environments.

The FTC accepted these data and arguments, and thus this tampon product is currently labeled flushable and biodegradable.

APPENDIX A

TEST METHODS

A.1 Screening Toilet Bowl Drainline Clearance Test

1.0 Purpose

This test is used to determine the relative clogging potential of a product as it passes through a toilet and household drainage system.

2.0 Scope

This test method evaluates the relative ease with which a product passes through a toilet and household drainage system. The test uses actual toilets and drainlines that are set to specific country plumbing codes. A positive result in this test would indicate the test product is compatible with household toilets and plumbing systems.

3.0 Principles

The test system consists of toilets and drainlines that are representative of those found in North America, Europe, and Asia. Flushable products are evaluated for their relative ease to evacuate the toilet bowl and trap, as well as to be transported through a simulated drainline plumbing system. This screening level test consists of trials of five product flushes each. The loading of the product is under heavy use (95th percentile user). This test condition is designed to be worst case to stress the system. Products that pass the screening level test do not need to be tested at the higher tier confirmatory evaluation.

4.0 Responsibilities

Researchers in the Global Flushability Lab or contract persons reporting to them are responsible for conducting this test.

5.0 Safe Practices

Immediately clean up any spilled water or simulated fecal material (SFM) from the test apparatus because it may cause a slipping hazard. Wear slip resistant shoes for stepping up on the test stand.

6.0 Procedure

6.1 Test System Design

The test system consists of two sets of toilets and drainlines. The first test stand has three toilets connected to a common 100 mm ID (4″) drainpipe. The total length of the drainline, depending on the position of the toilet on the test stand, should be > 60′ (18 m). The drainline includes four bends constructed of two 45° elbows. The line pressure of the water supply to the toilets is 20 ± 3 PSI. A standard 4″ ID (100 mm) Wye connector with 1/8 bend is inserted at the junction between the toilet drainline and the test drainpipe. The second test stand has four toilets with each connected to one of two 75 mm ID (3″) drainlines. The total length of the 75 mm drainline should be > 60′ (18 m), depending on the position of the toilet on the test stand. The slope of each of the drainlines is adjustable to meet the building codes of different regions. The required slope will be noted in the protocol and study records.

6.2 Functional Performance

Prior to conducting a test, the toilet should be checked for its functional performance. It is recommended that the ASME A112.19.6-1995 tests be used for this purpose. In particular, the 7.1.3.1 Ball Test or equivalent test should be used to assess the toilet's performance and the 7.1.6.1 Water Consumption and Hydraulic Characteristics Test or equivalent test should be used to assess the appropriate amount of water being flushed. During toilet flushing, the flush lever should not be held down, but released immediately. The tank volume controller should be varied until the appropriate flush volume is obtained (e.g., 1.6 gal/flush).

6.3 Experimental Design

The screening level test evaluates several trials of five flushes per trial. These trials may consist of the product only, product with another product, and/or product with simulated fecal material (SFM). The SFM is loaded at 160 ± 15 g per flush. This loading of SFM is based on the average amount of fecal matter disposed of per person per day. The product loading to the toilet is targeted for the 95th percentile user based on habits and practices data. This loading rate corresponds to < 1% failure rate on a per use basis.

6.4 General Flushing Instructions

1. Flush the toilet to be used to set the correct water level in the bowl and to clear any residue that may be in the drainline.
2. Place each piece of SFM or product into the toilet at 10-second intervals to allow time for the product to wet out.
3. Flush the toilet and observe the evacuation of the bowl.
4. Watch for the SFM and products to enter the drainline.
5. Enter data for each flush on the Attachment 2 Worksheet by following the tables in Attachment 3.
6. Continue as directed on the table until all five product flushes are completed.

If at any time in the flushing sequence, the product remains in the toilet bowl or trap after flushing, the toilet is either flushed again or the tissue is plunged into the drainline manually and the flushing sequence continues. After completion of each trial, the drainline will be cleared prior to beginning subsequent testing.

6.5 Concentration or Dosage Levels

The screening level test is typically run at the 90th percentile loading, based on P&G Habits and Practices data from each region. In the absence of H&P data, expected usage scenarios provided by Product Development can be used. The loading will be documented in the protocol or study records.

6.6 Sample Preparation

In the screening test, each strip of bath tissue is folded to the size of one sheet of tissue (approximately 4″ x 4″) before placing in the toilet. Folding the bath tissue has been shown in previous tests to be more conservative (i.e., harder to flush) than crumpling the tissue. Other products will be placed in the toilet according to their expected usage.

6.7 Type and Frequency of Tests, Analyses and Measurements

The following types of data will be recorded on the worksheets as instructed on Attachment 3:

1. Bowl/Trap clearance.
2. Distance the product travels on each flush.

6.8 Data Evaluation and Reporting

The data from each trial will be displayed on the following tables:

Bowl/Trap Clearance Data

Product	Clogged bowl (%)	Clogged trap (%)	Number of plunges needed to clear clogs	Average distance traveled per flush	Average number of flushes to clear drainline

6.9 Statistical Methods

Statistical methods, if used, will be described in the study records.

7.0 Documentation/Record Keeping

All records necessary to reconstruct the study and demonstrate adherence to established protocols will be maintained.

8.0 References

1. ANSI Standard A112.19.2M-1998
2. R. B. Martin, Plumbing Standards October 1989, page 17
3. ANSI A112.19.6-1995 Hydraulic Performance for Water Closets and Urinals
4. Japanese Industrial Standard A 5207 (1995)

5. Standard Sewerage Ordinance and Methods for the Examination of Flushability of Sanitary Products (Japan)

9.0 Attachments

Attachment 1—Preparation of SFM
Attachment 2—Worksheet
Attachment 3—How to take toilet and drainline measurements

ATTACHMENT 1

Preparation of Simulated Fecal Material

Purpose

Simulated Fecal Material (SFM) is used to simulate feces being flushed in a toilet.

Scope

Feclone is purchased from SiliClone Studios as a dry powder and must be mixed with water and placed in tubes to correctly mold the shape.

Responsibilities

Technician prepares and stores the Feclone and keeps track of the supply and notifies lab manager of the need to order more.

Schedule

The preparation of SFM is carried out on an as needed basis.

Safe Practices

1. Unplug the mixer and extruder while assembling or disassembling the equipment.
2. Always wear insulated gloves when carrying and pouring the boiling water.
3. Pour water slowly to avoid splashing.
4. Keep hands clear of mixer while it is running.
5. While loading SFM into the extruder, keep hands out of chute.
6. Immediately clean up any spills to prevent a slipping hazard.

Procedure

1. Heat 2900 ml of tap water to boiling in three 2 L flasks.
2. Slowly pour the water into the mixing bowl of the Hobart mixer.
3. Add one bag (approximately 900g) of Feclone BFPS-7 dry concentrate.
4. Add 20 ml of 1.3% Dawn solution (500 ml water and 6.5 ml of Ultra Dawn).
5. Mix on low for one minute.
6. Stop the mixer and scrape the sides of the bowl to mix in all the dry powder.

7. Increase the speed to medium and continue mixing an additional two minutes or until all dry powder is mixed in.
8. When the material is well mixed, transfer to the extruder pan.
9. Cut small squares of SFM and feed them down the chute.
10. Extrude the SFM into 50 ml disposable centrifuge tubes that have a small hole punched in the end of each.
11. Place screw caps on the end of each tube and place in the refrigerator if it will be used within the next three days. If the storage time will be longer than three days, store in the freezer to prevent fungal growth.
12. Before using, place in the water bath at 38°C.
13. After using the SFM, the tubes should be washed in a soapy Dawn (or a glassware cleaner) solution and scrubbed with a brush to remove all remaining SFM.
14. Soak the tubes in a 10% Clorox solution, rinse and air dry before refilling.

ATTACHMENT 2

Study No.________________ Test System____________ Date ________ Positions #1

Test Material ________________ Test Conditions ____________________

Trial # ____________________________ SFM________________

				Location in Drainline				
Flush #	**Bowl Loading**	**Remaining in Bowl**	**Remainin g in Trap**	**0-4 m u,p,1s,1e,2s**	**5-8 m 2s,2e,3s**	**9-12.5 m 3s,3e,4s,4e,5s**	**Clear Drainline**	**Note #**

Notes:

ATTACHMENT 3

Table A-1. Bowl and Trap Clearance

Condition	Record in column headed, "Remaining in bowl"	Record in column headed, "Remaining in trap"	Next step
The product hesitates in the bowl and water level raises before it all clears. (labored flush)	LF	0	Go on to the next product flush. [1]
Bowl is empty after flush and all solids are in the drainline.	0	0	Go on to the next product flush. [1]
Bowl is empty after flush but not all solids are in the drainline. The water level is normal.	0	List whatever is presumed to be in the trap. *Example: T1 for tissue from the 1st flush.*	Go on to the next product flush. [1]
Bowl is empty after flush but not all solids are in the drainline. The water level is high.	0	List whatever is presumed to be in the trap. *Example: T1 for tissue from the 1st flush.*	Plunge the product into the drainline. [1,2]
Product is still in the bowl after the flush.	List whatever is remaining in the bowl.	If all the solids can be accounted for between the bowl and drainline–0. Otherwise, list what is presumed to be in the trap.	Try an extra flush. If that doesn't clear the bowl, use the plunger to push the product into the drainline. [1,2]
Product is stuck in the trapway, visible in the bowl but no product is floating in the bowl.	0	List whatever is presumed to be in the trap. *Example: T1 for tissue from the 1st flush.*	Use the plunger to push the product into the drainline. [1,2]

1. Wait until the products have stopped moving and the water is only trickling out of the drainline to start the next product loading. Record the location of the solids in the drainline (see Table A-2).
2. When plunging, remember that one plunging incident consists of three quick plunges in succession. Record each plunging incident in the notes section. If the water level is low after plunging, reset with an extra flush. Each extra flush is recorded on a new line along with the location of the solids in the drainline.

Table A-2. Drainline Clearance (Refer to worksheet on the following page)

Flush #	Explanation of Drainline Data	Next step
1	The SFM and tissue cleared the toilet and ended up in the 1st straight pipe. The tissue traveled at least 2m but less than 3m	After the product has stopped moving in the drainline and the water slows to a trickle, start next loading.
2	S1T1 moved to 11m, the SFM from the 2nd flush (S2) stopped in 2E and the tissue (T2) stopped at 3m. S1T1 traveled to 11m.	After the product has stopped moving in the drainline and the water slows to a trickle, start next loading.
3	S3 cleared the toilet and stopped in 1E but all of the T3 tissue stayed in the trap. S2T2 traveled to 10m and S1T1 cleared the drainline.	Since the water level is normal, it may clear with just an extra flush.
4	The empty flush cleared the T3 tissue from the toilet and it traveled to 12m. S2T2 and S3 cleared the drainline.	Continue on with next loading as above.
5	The S4T4 loading hesitated before leaving the toilet but it cleared without intervention. S4T4 went to 5m and T3 cleared the drainline.	Continue on with next loading as above.
6	The S5 cleared the toilet and was in the 1st elbow. Only 1 strip of tissue was in the bowl and none was visible in the drainline, so 3 strips of T5 must be in the trap. S4T4 traveled to 11m.	Since the water level is normal, it may clear with just an extra flush.
7	The extra flush failed to clear the toilet but S5 moved to 4E and S4T4 cleared the drainline. TheT5 tissue was plunged into the drainline and it traveled to the 1st elbow (4m).	Empty flush to reset water level in bowl after plunging and also to move the tissue through the drainline.
8	The empty flush reset the water level land moved the T5 loading.	Continue empty flush to clear the drainline.
9	T5 clears the drainline after empty flush.	Trial is complete.

Study No. PEL02-XXX Test System 100mm /UK Date 07/17/02 Positions #1
Test Material Charmin LDT CLXXX Test Conditions 1.25% slope, no occlusion
Trial # SFM + 4x7 Charmin SFM 160g

				Location in Drainline				
Flush #	**Bowl loading**	**Remaining in bowl**	**Remaining in trap**	**0-4 m u,p,1s,1e,2s**	**5-8 M 2s,2e,3s**	**9-12.5 m 3s,3e,4s,4e,5s**	**Clear drainline**	**Note #**
1	S1T1	0	0	S1T1 (2m)				
2	S2T2	0	0	T2 (3m)	S2	S1T1 (11m)		
3	S3T3	0	T3	S3		S2T2(10m)	S1T1	
4	Empty flush	—	0			T3 (12m)	S2T2S3	
5	S4T4	LF	0		S4T4(5m)		T3	1
6	S5T5	T5d	T5abc	S5		S4T4(11m)		2
7	Extra Flush	T5d	T5abc			S5	S4T4	3
8	Empty flush	—	—			T5 (9m)	S5	4
9	Empty flush	—	—				T5	5

This worksheet is for a toilet in position #1 on the first test stand, which is connected to the 100 mm drainline. Use the appropriate worksheet corresponding to the position of the toilet and drainline in use.

m = meter if the location is between 2 meter markings, record as the lower number. Distance is only recorded for test product.

u = under the test stand

p = in the port (inverted T that is used as a drainline vent.)

s = straight pipe (the first straight pipe, between the port and the 1st elbow is designated 1s)

e = elbow (bend in the pipe constructed with two 45° elbows).

Notes:
1. Labored flush–tissue hesitated and water level went up before clearing.
2. T5 consists of four strips of tissue. Three strips are in the trap and one strip remains in the bowl after flushing.
3. Empty flush failed to clear the toilet.1XP (one plunging incident) After plunging, T5 was all in 1E (4m)
4. Empty flush to reset water level in the toilet bowl.
5. Empty flush to move the final loading through the drainline.

A.2 Column Settling Test

1.0 Purpose

This test is used to determine the rate and extent of settling of test material in a column of water that is at a depth similar to that of a typical septic tank.

2.0 Scope

This test evaluates the extent to which a test material is able to settle with other wastewater solids in a septic tank or wastewater treatment system. If a product does not settle in a septic tank, it may leave the tank with the effluent and cause problems in the drainage field. The column-settling test can quickly identify products that may not settle at an adequate rate to be retained in the septic tank.

3.0 Principles

The column-settling test was designed to simulate products settling in a septic tank, but can also be used to assess the settling of a test material in a grit chamber or a primary clarifier. The depth of the column is approximately the same depth as that of an average size septic tank, which is approximately 4′ (120 cm). The clear pipe used in this test provides a view of the entire depth of the column to observe the settling behavior of the product. The column can be filled with tap water or raw wastewater. The rate of descent is then recorded.

4.0 Responsibilities

Researchers in the Global Flushability Lab or contract persons reporting to them will be responsible for conducting this test.

5.0 Safe Practices

This procedure may involve the use of wastewater from a municipal sewage treatment plant or septage from a septic tank. Consequently, individuals performing this test may be exposed to microbiological agents that are potentially hazardous. Follow normal safe handling precautions when handling biologically active samples and any chemicals.

6.0 Procedure

6.1 Experimental Design

The settling column is a transparent plastic pipe that is mounted vertically on a test stand. The depth of the pipe is 152 cm and the inside diameter is 20 cm. There is a removable plug in the bottom of the column so that water can be drained. The column holds approximately 45 L of water. This test is typically combined with a bowl and drainline clearance test. As the product clears the drainline, it is collected in a beaker with approximately 1 L of water. The water and product are poured into the test column that has been filled to the 3 cm mark with water. The timer is started when the sample enters the column of water. The time measured is the length of time it takes for the sample to settle to the 115 cm mark near the bottom of the column. If a sample remains floating for five minutes, 1 L of water is added to the column. If the product still remains floating, additional water is added at 0.5, 1, 2 and 23 hours. If multiple additions of water are needed, 1 L of water should be drained from the column before adding the next liter in the top of the column. If at the end of 24 hours the product has not settled, the test is terminated.

6.2 Concentration or Dosage Levels

The amount of product added to the test column is typically the average amount of product that would be flushed, based on habits and practices data.

6.3 Sample Preparation

Product to be added will be prepared in the same manner as used in the flushing test if the product is to be flushed before settling. Any other sample preparation will be detailed in the protocol or study records.

6.4 Type and Frequency of Tests, Analyses and Measurements

The column-settling test consists of five replicates. Each replicate is flushed and/or dosed separately. The amount of time required for the product to settle to the 115 cm mark is measured.

6.5 Data Reporting

The average (± std dev) settling time of the five replicate samples is reported.

6.6 Proposed Statistical Methods

Statistical methods if used will be described in the study records.

7.0 Documentation/Record Keeping

All records necessary to reconstruct the study and demonstrate adherence to established protocols will be maintained.

8.0 References

None

9.0 Attachments

None

A.3 Aerobic Disintegration (Shake Flask) Test

1.0 Purpose

This test is used to assess the fate of a flushable product during transport through a sewer collection system.

2.0 Scope

This test method is used as a rapid screening tool to assess the fate of flushable products during transport through a sewer collection system. In particular, it assesses the rate and extent of disintegration of a test material in the presence of raw wastewater. A positive result of this test would indicate that the flushable product would not be recognizable from other wastewater solids at the influent of a wastewater treatment plant.

3.0 Principles

This test method is similar to other shake-flask die-away biodegradation test methods (e.g., ASTM, U. S. EPA), except that in this method raw wastewater is used—as opposed to river water and activated sludge—and whole product is used–as opposed to dissolved chemicals. Moreover, the shaking in these other tests is to ensure a sufficient amount of oxygen is present for biodegradation, whereas in this test it is to simulate the physical forces acting to disintegrate the product during sewer conveyance. In this test, whole product is placed in a 3 L flask containing either 1 L of wastewater or tap water and then is shaken on a rotary shaker table. Periodically, flasks will be removed and their content passed through a series of sieves. The various size fraction retained on the sieves are weighed and the rate and extent of disintegration is determined.

4.0 Responsibilities

Researchers in the Global Flushability Lab or contract persons reporting to them are responsible for conducting this test.

5.0 Safe Practices

This procedure involves the use of raw wastewater from a municipal wastewater treatment plant. Consequently, individuals performing this test may be exposed to microbiological agents that are potentially hazardous. Follow normal safe handling precautions when handling biologically active samples and any chemicals.

6.0 Procedure

6.1 Materials and Apparatus

The following materials are needed to perform this test.

1. Triple Baffled Fernbach Erlenmeyer Flasks (2800 mL), Bellco Glass Inc., Catalog number 2551-02800.
2. Orbital Floor Shaker with 2" orbit, 150 rpms. The platform for the shaker needs clamps to be able to accommodate a bottom flask diameter of 205 mm.
3. USA Standard Testing Sieves: 5/16 in (8mm), #5 (4mm), #10 (2mm), #18 (1mm).
4. Drying oven capable of maintaining a temperature of 40 ± 3°C for plastic test materials and capable of maintaining a temperature of 103 ± 3°C for non-plastic test materials.
5. Whatman #41 filter paper, 12.5 cm circles.
6. Shower head and hose assembly with rubber fitting for sink faucet.
7. Aluminum weighing dishes, two sizes: 43mm and 57mm.
8. Forceps.
9. Commercial paint brushes from hardware store.

6.2 Reference Material

A standard reference material (referred to as "control") is desirable to gauge the disintegration potential of the run. The recommended reference material is a Whatman Ashless Filter Paper #41, which should be folded and fluted before introducing it to the test vessel. This particular filter paper has a medium wet-strength, which allows it to slowly disintegrate over the testing period.

6.3 Field Sampling

Raw wastewater is collected from a municipal wastewater treatment plant (WWTP) upstream of any internal plant recycles and prior to any screening or grinding processes. The wastewater should be pre-screened through a 1.0 mm USA Standard Testing Sieve prior to dispensing into the test flasks.

6.4 Analytical Measurements

The following analytical measurements are recommended on the pre-screened wastewater, but not necessary.

1. Suspended solids
2. Chemical Oxygen Demand
3. pH

6.5 Test Initiation

For screening level testing, each test product is run in triplicate to accommodate three (3) predetermined destructive sampling time points. For confirmation level testing, each time point is run in triplicate. Each flask will contain either 1 L of pre-screened wastewater (see field sampling section above) or tap water and 3–5 g of the test product. Each test product should be pre-weighed in triplicate on a four-place balance and these weights recorded in the laboratory notebook for final percent disintegration calculations.

Control flasks with the reference material are also run in triplicate to accommodate three (3) destructive sampling time points. Each flask also contains either 1 L of pre-screened wastewater or tap water and 3–5 g of fluted/folded Whatman #41 filter paper (generally about three-four circles). The control product should be pre-weighed in triplicate on a four-place balance and these weights recorded in the laboratory notebook for final percent disintegration calculations.

Measure 1 L of wastewater or tap water into each of the Fernbach flasks and place them on the rotary shaker table. Add 3–5 g of test product to the flasks. Turn on rotary shaker and shake flasks at 150 rpms. Visual observations of the test product and the control product should be recorded after one hour. The recommended sampling times for most products are at six hours, 24 hours and 48 hours.

6.6 Test Termination

At the designated destructive sampling points, remove a flask from each set of products being tested and from the control set. Pour the entire content of a single flask through a nest of sieves arranged from top to bottom in the following order: 8 mm, 4 mm, 2 mm and 1 mm. With a handheld showerhead spray nozzle, gently rinse the material through the nested sieves, being careful not to force passage of the retained material through the next smaller sieve. After several minutes of rinsing, remove the top sieve and continue to rinse the next smaller sieve, still nested, for a few minutes. Again, be careful not to force passage of retained material to the next smaller sieve. Continue this rinsing process until all of the sieves have been rinsed. After rinsing is complete, remove the retained material from each of the screens by inverting the screen onto a smaller screen and rinsing the retained material in the reverse direction. This procedure is described in detail below. Transfer the content from each screen to a separate, labeled aluminum pan. Place the pan in the oven overnight at 103 ± 3°C (or other appropriate temperature) until the sample is dry. Continue this procedure at each designated sampling time point until all the test products are sampled. After all the samples are dry, weigh the materials from each of the retained fractions and calculate the percentage of disintegration based on the initial starting weight of the test material.

6.7 Sieve Sampling

Begin removing material from the smallest sieve (1.0 mm) first. Remove the material retained on the smallest screen and transfer it to an aluminum pan. Take the next larger sieve (2.0 mm) and invert the screen on top of the smallest (1.0 mm) sieve. Rinse the retained material from the larger screen onto the smaller one, then collect the material from the smallest sieve and transfer it to an aluminum pan. Repeat this procedure for the remaining sieves, continuing to transfer the content of the larger sieves to the smallest (1.0 mm) sieve and ultimately to an aluminum pan for drying.

In products that contain a lot of pulp (i.e. bath tissue), it is recommended to pour a small portion of the contents of the flask through the nesting of sieves and rinse the material for a few minutes before adding the next portion. This allows the pulp to better separate on the sieves (especially on the smaller grid sieves) and not be susceptible to "tangling," which adds to unnecessary retention of material on the sieves.

6.8 Calculation

Calculate the percent of disintegrated test product using the following equation:

$$\frac{\text{Dried weight of retained fraction on sieve (gm)}}{\text{Initial weight of sample (gm)}} \times 100 = \%\ \text{of initial added mass}$$

7.0 Documentation/Record Keeping

All records necessary to reconstruct the study and demonstrate adherence to established protocols will be maintained.

8.0 References

1. ASTM E 1279—89 (Reapproved 1995) Standard Test Method for Biodegradation by a Shake-Flask Die-Away Method.
2. U.S. EPA OPPTS 835.3170 (April 1996) Fate, Transport and Transformation Test Guidelines, Shake Flask Die-Away Test.

9.0 Attachments

None

A.4 Anaerobic Disintegration Test

1.0 Purpose

This test is used to determine the rate and extent of disintegration of a test material under anaerobic conditions.

2.0 Scope

This test evaluates the extent to which an organic material undergoes biological disintegration in an anaerobic environment. A high rate of disintegration provides evidence that the test material will not be visible in anaerobic digester sludge or septic tank septage.

3.0 Principles

This is a 28-day test in which the test material is exposed to biologically active anaerobic digester sludge. The disintegration of the test material is measured by weight loss over time. The conditions are optimized for this test (temperature and level of solids) to maximize the rate at which the test material will degrade. The results of this test are indicative of what happens to a product in the sludge layer of a septic tank or in an anaerobic digester at a municipal wastewater treatment plant. This test was modeled after a test used by the National Sanitation Foundation (NSF) to determine how well flushable products degrade in an anaerobic environment, such as a septic tank or in an anaerobic digester.

4.0 Responsibilities

Researchers in the Global Flushability Lab or contract persons reporting to them are responsible for conducting this test.

5.0 Safe Practices

This procedure involves the use of biologically active anaerobic digester sludge. Consequently, individuals performing this test may be exposed to microbiological agents that are potentially hazardous. Follow normal safety precautions when handling biologically active samples or any chemicals.

6.0 Procedure

6.1 Experimental Design

6.1.1 Test Design

The test materials and control products are dried, weighed and added to 2 L reactor bottles containing 1.5 L of anaerobic digester sludge. A minimum of three reactor bottles is prepared for each test material. The reactor bottles are then placed in an incubator set at 35°C. On sampling days (typically days 7, 14 and 28), a reactor bottle for each test material is harvested. The content of each reactor bottle is poured onto a 1 mm sieve. The sludge is then gently rinsed off of the test material residues. The material that is retained on the sieve is then air dried at 100°C (or another temperature depending on the melting point of the test material) overnight in an oven. The dried and desiccated samples are then weighed and the percent o weight loss is calculated.

6.1.2 Anaerobic Digester Sludge

The anaerobic digester sludge is obtained from an anaerobic digester at a local municipal wastewater treatment plant that receives predominantly domestic wastewater. The sludge is screened through a No. 18 mesh screen (1 mm openings) before use. The sieved sludge must meet the following criteria for the test to be valid:

Parameter	Criteria
Total solids	>10,000 mg/L
PH	6.5 to 8.0

In addition, the control product (a tampon) must have > 95% weight loss through a 1 mm opening during the 28-day testing period.

6.2 Test Procedure

1. Sieve the sludge through a No. 18 mesh screen to remove any solids greater than 1 mm in size. If the Sweco Separator is used for screening, use two screening decks with the coarser screen on top. This courser screen helps keep the 1 mm screen from blinding. (See the Sweco SOP)
2. Pour the screened sludge into a mixing barrel and cover with a plastic bag.
3. Insert the tubing from the nitrogen cylinder through a small hole in the bag and secure it to the barrel with a clamp or duct tape.
4. Turn the gas on at a low flow rate so that it blankets the surface of the sludge in the barrel.
5. Insert the stirrer blade and shaft through another small opening in the bag and lower the stirrer down to near the bottom of the barrel. When the gas has had several minutes to fill the headspace in the barrel, turn the mixer on.
6. Use a peristaltic pump to transfer 1.5 L of sludge into the 2 L reactor bottles. Place a one-hole stopper on each bottle as it is filled.
7. Label each bottle and add the pre-weighed sample.
8. Invert the bottle a couple of times to get the sample into the sludge.
9. Place the bottles in an incubator set at 35°C.

10. On sampling days (typically days 7, 14, and 28), the content of each bottle is poured through a No.18 sieve (1 mm opening). Rinse the remaining sludge solids through the sieve with tap water. Any sample residue remaining on the sieve is placed in a dish and dried in the oven overnight. The standard drying temperature is 103 ± 3°C, however, this temperature may vary depending on the properties of the test material. The drying temperature will be noted in the protocol or study record. Photographs should be taken of the sample before drying.
11. Samples are desiccated and then weighed to determine percent of weight loss of the test material.

6.3 Concentration or Dosage Levels

The amount of test material dosed to each reactor bottle will be given in the protocol or study record. Three control bottles will be prepared the same way as the sample reactor bottles and dosed with one pre-weighed tampon each.

6.4 Sample Preparation

Samples are weighed before dosing to the reactor bottles. Any other sample preparation such as drying will be described in the protocol or study record.

6.5 Type and Frequency of Tests, Analyses and Measurements

Typically, there are three sampling days per test, i.e., days 7, 14 and 28. These sampling days can be changed depending on the expected disintegration time of the test materials. The number of replicates, test duration and sampling days will be documented in the protocol or study record.

6.6 Data Reporting

All data will be captured in the study record and a report will be prepared that summarizes the data. The data to be included in the study report are the dried weights of the test material for each time point and observations concerning the physical condition of the sample. Photographs of the test material may also be included in the study report.

6.7 Statistical Methods

Statistical methods if used will be described in the study records.

7.0 Documentation/Record Keeping

All records necessary to reconstruct the study and demonstrate adherence to established protocols will be maintained.

8.0 References

None

9.0 Attachments

None

A.5 Aerobic Biodegradation (Electrolytic Respirometry) Test

1.0 Purpose

This test is used to determine the rate and extent of biodegradation of a test material in an aerobic environment.

2.0 Scope

This test evaluates the extent to which an organic material undergoes aerobic biodegradation. A positive result in this test indicates that the test materials will biodegrade in a home aerobic treatment system, an activated sludge treatment system and in the soil environment. This test method is applicable for test concentrations that are not inhibitory to aerobic bacteria.

3.0 Principles

An electrolytic respirometer (Coordinated Environmental Services, Ltd.) is used to measure the extent and rate at which an organic substance undergoes biodegradation. This instrument continuously and automatically measures the amount of O_2 consumed by microbes as well as the amount of CO_2 produced. Each system consists of 20 sensor heads (Figure A-1) housed in a water bath. A single chiller and computer controller can operate as many as five systems. The test material, media and inoculum are incubated in a flask connected to a sensor head that contains a CO_2 probe, a Cu_2SO_4 reservoir, and a pressure transducer. As CO_2 is liberated from the media, it is absorbed by the alkali trap within the CO_2 probe. The change in conductivity of the base solution is automatically recorded. As the CO_2 is removed from the test vessel, the pressure in the headspace is reduced. This reduction is detected by the pressure transducer, which triggers the electrolysis of the Cu_2SO_4 and liberates free O_2. The amount of O_2 needed to restore this pressure is proportional to the magnitude and duration of the electrolyzing current.

4.0 Responsibilities

Researchers in the Environmental Fate Testing Lab will be responsible for conducting this test.

5.0 Safe Practices

This procedure involves the use of activated sludge from a municipal sewage treatment plant treating predominately domestic wastewater. Consequently, individuals performing this test may be exposed to microbiological agents that are potentially hazardous. Follow normal safety precautions when handling biologically active samples or any chemicals.

6.0 Procedure

6.1 Salts Media Preparation

The media contains the following solutions per liter of high quality water (HQW):

Magnesium sulfate 2.25% aqueous solution	1 mL
Calcium chloride 2.75% solution	1 mL
Ferric chloride solution 0.025% solution	1 mL
Phosphate buffer, pH 7.0 (see below)	10 mL

Phosphate buffer solution (amounts given are per liter of HQW):

Potassium dihydrogen orthophosphate, KH_2PO_4	8.5 g
Dipotassium hydrogen orthophosphate, K_2HPO_4	21.75 g
Disodium hydrogen orthophosphate dihydrate, $Na_2HPO_4 \cdot 2H_20$	33.40 g
Ammonium chloride, NH_4Cl	0.5 g

NOTE: The pH of the media is adjusted from ~7.4 to 7.0 prior to use with 2N HCl.

6.2 Inoculum Preparation

The activated sludge inoculum is obtained from one of the following sources:

1. Wastewater treatment plant receiving predominately domestic sewage.
2. Porous pot unit that has been acclimated with the test substance.
3. Semi-Continuous Activated Sludge unit that has been acclimated with the test substance.

The inoculum is prepared using one of the described methods.

1. The activated sludge is homogenized at room temperature for two minutes (±30 sec) at medium speed in a Waring blender or equivalent. This homogenized sample is allowed to settle for 30–60 minutes before the supernatant is decanted. The supernatant is then filtered through glass wool to remove any remaining sludge solids. This inoculum may be added at a predetermined concentration (typically 1% v/v) to the salts media and aerated overnight with CO_2 free air to reduce the background carbon level or to the sample bottles individually. Total suspended solids analysis may be performed on the prepared inoculum, if desired.
2. Adjust the activated sludge solids to a target concentration of 3000 mg/L (using total suspended solids analysis). Centrifuge the sludge for 10 minutes (5000 rpm). Decant the supernatant and reconstitute the sludge solids with an equal volume of the salts media. Repeat wash procedure, if desired. Homogenize the sludge as described above. Do not

filter the inoculum. The inoculum may be added to the media as described above or to the sample bottles individually. *Adding 1% v/v (10 mL/L) will achieve the maximum level of sludge solids allowed for the OECD Guidelines 301B Test Method (30 mg/L).

6.3 Total Organic Carbon (TOC) Analysis

TOC analysis should be conducted on all test material stock solutions to determine the amount of carbon added to each test sample (refer to BSTS SOP [Volume II Section IV-61]). If the test material is insoluble, the percentage of carbon should be determined by another analytical method or based on the structure.

6.4 Test Preparation (one day prior to test initiation)

The water bath is filled to 2/3 capacity, the temperature set at 22 ± 0.5°C and the recirculating pump started. An algaecide may be added at approximately weekly intervals throughout the test to keep the water free of alga growth.

The $CuSO_4$ reservoirs can be filled with 40–50 mL of 25% $CuSO_4$ solution. This solution can be used for several experiments, as long as the solution retains a strong blue color. Verify that each oxygen cell contains a pair of electrode wires (anode and cathode). The cathode wire (thicker of the two) should be replaced if it contains more than a 10 mm Cu^{++} build up. The anode needs to be replaced only if broken. Check to make sure that both the empty reference bottle and oxygen cell container are sealed tightly in the sensor head.

6.5 Preparation and Set up of Base Traps

The 1% NaOH solution for the CO_2 probes is prepared the day before the experiment begins. Once the base is well mixed, it is degassed via filtration using a glass fiber filter. The base (18 mL) is dispensed accurately into the C-caps and then is gently attached to the probe. Use care to avoid air bubbles in the base. Attach a bottle (100 mL may be used) to the end of the probe so it remains airtight. *NOTE: The filling C-cap probes should be filled in as quickly as possible in order to minimize absorption of CO_2 from the air to the base.* Immerse the sensor assembly into the water bath; the water level should be at least halfway up the sensor head. After the probes have been immersed, the instrument is started by selecting the green traffic light on the main screen of the respirometer software. Select all channels to be used in the experiment. Enter the log rate (sampling interval) and initiation time (stabilization time). One hour should be sufficient for both parameters for monitoring the probes. Click "OK" to activate the selected channels. The probes will be monitored overnight while the base equilibrates. The channel numbers should turn black for the initiation phase and then green for the testing phase.

If any of the probes are not being used, add fresh base to the caps and store them in the same manner. The probes should not be allowed to dry out. If the probes are allowed to dry out, they will have to be reconditioned by soaking in fresh base for a day or two before they can be used again.

6.6 Test Initiation

1. The following day, stop the instrument by selecting the red stop light at the top of the main menu. The channels numbers should turn red. Select "Acquire" to ensure all the data files have been updated and then select the "Graph" button. Look at the readings for

both O_2 and CO_2 to determine if the base has equilibrated and if the probes are functioning properly. The readings may jump around initially but should settle down after several hours. If the system appears stable, dispense 1 L of the aerated media into each Pyrex bottle and add a stir bar. The inoculum (1% v/v) should also be added at this time, if it has not been premixed with the media. The test substance is added to each test bottle to achieve the concentration specified in the protocol or determined by the Study Director. For solid substrates, 15–20 mg-C/L is typically added to the test vessels to ensure a good signal to noise ratio. A reference material (sodium benzoate) is also included with every test. This material is used to verify that the inoculum is active and the instrument is operating properly. The blank bottles receive no test substance (media and inoculum only). The reference and blank samples are run in triplicate.

2. Exchange the sample bottles for the empty bottles that were used to seal the probes and adjust the stirring mechanism to ~90% full power.
3. Immerse the sensor assembly into the water bath so the water level is at least halfway up the sensor head or high enough so the sensor assembly is completely immersed throughout the test period. Start the respirometer control program by selecting all channels being used and clicking on the green traffic light. Enter the log rate (sampling interval) and initiation time (stabilization time), then click "OK." The selected channels will be activated. Suggested parameters include a sampling interval of 720 minutes (12 hours) and initiation time of one hour.

6.7 Test Period

This test is typically conducted for approximately 28 days, but may run longer depending on the observed data. If the CO_2 production in either the test bottle or the positive control (sodium benzoate or other) does not reach a plateau as determined by the Study Director, contact the sponsor before terminating the study.

6.8 Test Termination

When the CO_2 production reaches a plateau and/or the study is terminated by the direction of the Study Director and/or Sponsor, the final data is acquired and the instrument is stopped by selecting the red traffic light. This will stop the data collection. The sample bottles are removed from the sensor head and sealed until analyzed. The soluble organic carbon remaining in each bottle is determined by TOC analysis.

When the CO_2 production is still increasing at test termination, sample acidification may be desired. If the samples are to be acidified, the sample bottles must be equipped with a sampling port and the following procedure should be used. Pause the system by clicking on the yellow traffic light. Add 1 mL of 50% HCl to each bottle to drive off the inorganic carbonate. A syringe can be used to inject the acid through the septum of the sampling port. The system is restarted by selecting the yellow/green resume button. The system continues to operate for one or more days. This allows for the inorganic carbon to be absorbed in the base trap.

(Optional) The base may be discarded and the C-Cap rinsed and gently cleaned with water. Refill the cap with fresh base and reconnect to probe. Seal the probe using an empty bottle (100 mL or other). Parafilm may also be used. However, fresh base should always be used when starting a new test.

6.9 Calculations

During the course of the test, a study report can be generated to determine either the percent of CO_2 produced or O_2 consumed by the microbial population. Prior to generating this report, the TCO_2 and/or $THOD_{NH3}$ must be determined in addition to the amount (mg) of test material added to each flask. For mixtures, the TCO_2 of the total test material is a weighted average of the TCO_2 of the individual components. The average CO_2 production from the sample blanks is subtracted from each test sample prior to determining the final CO_2 production value.

6.10 Calculation of Theoretical CO_2 (TCO_2)

TCO_2 = mg CO_2/mg of test material

= # of carbons in test material x mol. wt. of CO_2 / mol. wt. of test material

or

If the formula for a test material is not provided:

mg CO_2 for a given TOC level = measured TOC (mg/L) * 3.667 (mol. wt. CO_2/mol. wt. C)

6.11 Calculation of Theoretical Oxygen Demand ($THOD_{NH3}$)

For the compound: $C_cH_hCl_{cl}N_nNa_{na}O_oP_pS_s$, the THOD without nitrification is calculated by the following equation.

$THOD_{NH3}$ = mg O_2/mg of test substance

$$= \frac{16[2c + \frac{1}{2}(h - cl - 3n) + 3s + \frac{5}{2}p + \frac{1}{2}na - o]}{mol.wt.}$$

This calculation implies that C is mineralized to CO_2, H to H_2O, P to P_2O_5 and Na to Na_2. Halogen is eliminated as hydrogen halide and nitrogen as ammonia.

The TOC analysis of the stock solution will provide a measured carbon level. Based on this level, determine the amount of test material that was dosed to the test vessel. This mg amount is the value used for the channel parameter section of the respirometer software. The data can be imported to an Excel workbook for analysis, if desired.

7.0 Documentation/Record Keeping

All records necessary to reconstruct the study and demonstrate adherence to established protocols will be maintained.

8.0 References

OECD Guideline 301B for Ready Aerobic Biodegradability—Gas Production from a Diluted Activated Sludge.

9.0 Attachments

Figure A-1—Respirometry unit

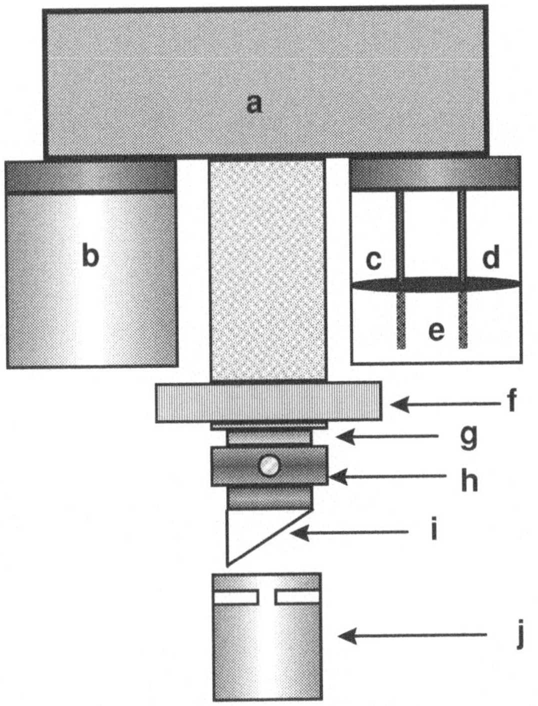

a	Sensor head
b	Air chamber
c, d	Anode & cathode
e	Saturated $CuSO_4$
f	Sample bottle cap
g, h	O-ring & vent
i	Electrode
j	Cuvette for electrolyte

Figure A-1. Respirometry unit

A.6 Anaerobic Biodegradation (Respirometry) Test

1.0 Purpose

This test is used to determine the rate and extent of biodegradation of a test material in an anaerobic environment.

2.0 Scope

This test evaluates the extent to which an organic material undergoes anaerobic biodegradation. A positive result in this test indicates that the test materials will biodegrade in anaerobic environments. The procedure for determining the conversion of organic matter to gas is described. This method is applicable to test concentrations that are not inhibitory to anaerobic bacteria.

3.0 Principles

The test material is incubated in a mineral media with a dilute concentration of inoculum (anaerobic digester sludge). Preparation of the media and inoculum is performed under a nitrogen blanket to ensure anoxic conditions. The test vessels (160-mL serum bottles) are sealed with a rubber stopper and gas production is measured using a pressure transducer (Omega Engineering, PX236). The pressure transducer response is continuously and automatically monitored with an anaerobic assay unit (Technical Automation Services Corporation). The millivolt (mv) readings are digitally collected using Lab logic software (National Instruments). The data are easily imported into Excel for further manipulation. The measured gas production from a blank containing no test material and the response from the bottles containing the test material is used to determine the extent of biodegradation. The test is typically run for 56 days or until the gas production reaches a plateau. This method was developed from standard test methods for organic chemicals (see *Reference* section).

4.0 Responsibilities

Researchers in the Environmental Fate Laboratory are responsible for conducting this test.

5.0 Safe Practices

This procedure involves the use of anaerobic digester sludge from a municipal sewage treatment plant treating predominately domestic wastewater. Consequently, individuals performing this test may be exposed to microbiological agents that are potentially hazardous. Follow normal safety precautions when handling biologically active samples or any chemicals.

6.0 Procedure

6.1 Reagents, Equipment, and Materials

6.1.1 Reagents (VWR, Fisher or equivalent)

Anhydrous Potassium dihydrogenphosphate (KH_2PO_4)
Disodium hydrogen dodecahydrate ($Na_2HPO_4 \cdot 12H_20$)
Ammonium chloride, NH_4Cl
Calcium chloride dihydrate ($CaCl_2 \cdot 2H_20$)
Magnesium chloride hexahydrate ($MgCl_2 \cdot 6H_20$)
Iron (II) chloride tetrahydrate, ($FeCl_2 \cdot 4H_20$)
Resazurin (oxygen indicator)
Sodium sulfide nonhydrate (Na_2S)
Sodium benzoate

6.1.2 Equipment

Technical Automation Services Corp. (TASC) anaerobic assay unit:
- Incubator oven (maintain 35°C) with capacity for 32 serum bottles
- IBM computer system loaded with Labview 6B Data Logger v1.0
- 32 pressure to analog signal transducer assemblies

IPS connected to computer system and backup power source for all electronics
Heating mantle (3 L size) and controller
3 L round bottom boiling flask (with short neck)
TOC analyzer
Water bath (35°C setting, 32 serum bottle capacity)
Drying oven (105°C)
Centrifuge (250 ml tubes)
Tabletop centrifuge

6.1.3 Materials

160 mL serum bottles (Wheaton)
Butyl rubber stopper (Bellco Glass, 2048-11800)
Aluminum crimp seal (Bellco Glass)
Crimper
Filtration apparatus
1 L amber bottle
250 mL centrifuge bottles
Analytical balance
Stir bars
One regular size stir plate
Gas tight syringes, 10-50 mL
Ring stand
Gas lock with stopper attached to fit 0.5 L bottle (used in wine making)

Clamps to hold tubing, thermometer
Thermometer (read 100°C)

6.2 Test Setup (day prior to test initiation)

It is recommended that the media and inoculum be prepared prior to the day of the test initiation.

6.2.1 Inoculum Collection

Collect the sludge solids from an anaerobic digester at a WWTP that receives predominately domestic wastewater. Use a wide neck polyethylene bottle or other non-glass container. Fill the container within 5–7 cm of the top and seal tightly. 4 L of digester sludge is more than enough for a single test. In the laboratory, flood the headspace of the container with nitrogen using one of the prepared sparge lines (see *Attachment 1—Preparation of Dispensing Solution and Sparge Lines*).

6.2.2 Media Preparation

See *Attachment 1—Preparation of Dispensing Station and Sparge Lines* before continuing with media preparation.

A media volume of 2.5 L is needed to conduct a 32-sample test. The media is prepared using the chemicals below in a boiling flask. Once prepared, this solution should turn blue after the addition of resazurin. Attachment 2 contains a form to be used when preparing anaerobic media.

Chemical	Amount per liter of solution
Anhydrous Potassium dihydrogenphosphate (KH_2PO_4)	0.27 g
Disodium hydrogen dodecahydrate ($Na_2HPO_4 \cdot 12H_20$)	1.12 g
Ammonium chloride (NH_4Cl)	0.53 g
Calcium chloride dihydrate ($CaCl_2 \cdot 2H_20$)	0.075g
Magnesium chloride hexahydrate ($MgCl_2 \cdot 6H_20$)	0.10 g
Iron (II) chloride tetrahydrate, ($FeCl_2 \cdot 4H_20$)	0.02g
*Resazurin (oxygen indicator)	0.001g
Na_2S	0.03g
Boiling Chips	
Large stir bar	
Water	1 L

6.2.3 Inoculum Preparation

The second sparge line is used to keep the sludge solids under a reducing environment during inoculum preparation. This procedure is described in *Attachment 2* on the "Anaerobic Inoculum Preparation Form."

6.2.4 Test Compound and Inoculum Concentration

The concentration for cellulosic base materials is typically dosed at 200–300 mg/L organic carbon. A typical range of 1–3 g/L inoculum solids is use to deliver the best signal to noise ratio (s/n). Table A-3 illustrates the s/n ratio based on inoculum level, where an s/n of > 3 is considered optimal. Test materials should be run in replicates of 3–5. TOC analysis should be conducted on all stock solutions, if possible, to measure the amount of carbon added.

Table A-3. Predicted signal to noise ratio based on inoculum concentration

	Inoculum (%, m/m)		
mg carbon	3%	2%	1%
400	9.7	12.8	23.9
200	5.3	6.8	12.3
100	3.1	3.8	6.4
50	2.0	2.3	3.6
25	1.0	1.2	1.8

6.3 Test Initiation (day 0 of test period)

The volume of test material added to the serum bottles should be < 5% of the total volume to avoid preparing an anoxic stock solution. A reference substance should be included in each test. Sodium benzoate, phenol, polyethylene glycol 400 or bath tissue is recommended. If the test compound is a solid or insoluble, it may be added to the serum bottle prior to addition of the media, however, a drop or two of water should be added to wet lightweight samples or the N_2 sparge will blow them out of the bottle.

Complete the "Anaerobic Respirometer Test Design Sheet" in *Attachment 2* with the calculated volumes of test chemical, inoculum, and media to be added for each treatment. Make sure the final total volume (typically 50 mL) is equal for all the treatments. Remember to adjust the media volume of the blank to make up for the lack of test material addition.

6.3.1 Dispensing Inoculated Media

Remove the rubber stopper from the flask and replace it with the opened sparge line. Using the Wheaton automatic dispenser, dispense a sample of media for pH and IC measurement. Measure the pH. If it is outside the range of 7.0 + 0.2, adjust it using as small a volume of acid/base as possible. Adjust the setting for sample volume. The remainder of the settings are based on user preference. Recommendations are as follows: 25 second delay, 3 mm tube size, 32 dose count (or number of samples to prepare), 125 rpm, 5 second ramp time and set on automatic mode. Drop the tubing line into the media so that the glass sinker rests near the flask bottom and turn on the stir plate. Fill water bath and set at 35°C. To fill the serum bottles, place the combination sparge/dispensing line over the top of the serum bottle and press "Start." When the media has been dispensed, carefully remove the sparge line as the rubber stopper is inserted into the bottle, and place the line into the next serum bottle. The tip of the sparge needle should be in the bottle until the stopper is put in place. Wetting the stoppers makes them easier to insert. The filled serum bottles are stored in the 35°C water bath. Add the metal crimp seal to each sample prior to connecting the bottles to the anaerobic system.

6.3.2 Incubating & Connecting Pressure Transducer to Samples

Make sure the pressure transducer valves are clear by opening the valve (arrow of handle facing away from pressure transducer) and placing the injection port of the valve up against an air source to blow any debris away from the needle. Close the valve (arrow of the handle is pointed towards the pressure transducer). Carefully insert the needle of the valve into the rubber stopper as far as it will go without forcing it. Once the pressure transducers have been inserted in all the serum bottles, vent the bottles by quickly opening and closing the valves. Put the bottles in the metal sample trays. Connect the samples to the instrument by hooking the four wire pins of the pressure transducer to the connector box. Close the oven door when all the samples have been connected.

6.3.3 Initiation of Test Using Software

Make sure all of the channels to be used are active and then set the sampling interval (data logging interval). Any channels that reads 0 to < 10 mv at start up requires no further action. If a channel reading is out of this range, connect a gastight syringe to the luer fitting on the valve and add ~5 mL of gas to the sample. If the value is < 0, remove 5 mL of gas from the sample until the reading is < 10 mv.

6.3.4 Test Termination

When the gas production reaches a plateau and/or the study is terminated by the direction of the Study Director and/or Sponsor, the final data is acquired and instrument is stopped. The samples are immediately opened and a sample of the liquid is analyzed for inorganic carbon (IC) analysis. This test is typically conducted for 56 days, but may run shorter or longer depending on the observed data.

6.4 Calculations

The following steps outline the procedure for setting up an Excel Spreadsheet and performing the necessary calculations:

1. Using Excel, create a workbook with separate sheets designated for each chemical, one sheet for the raw data and one for the blank. Download the raw data by selecting the data file and opening it as a WordPad file. Copy the contents into the raw data sheet of the Excel file. From this worksheet, cut and paste the data and sampling times of each chemical into its designated worksheet.
2. Typically, time 0 would be designated as 24 hours after the test is initiated. For the blank treatment, zero the data by subtracting out the millivolt reading at time 0 from all of the data points of each channel. Determine the mean value of the blanks. Any time points before the time 0 point can be deleted.
3. For the test substance, zero the data using the same time point as described in Step 2 and subtract the mean blank value from the appropriate time point. The data is now "blank and zero corrected" and in mv units.
4. To convert the data from mv to mL of gas, use the data generated from the linear regression analysis of the methane gas calibration curve (see *Attachment 1—Values Derived from Calibration*). Solve for "x" using the "y" intercept and the "b" slope that

were calculated by Tablecurve or an equivalent software package. The data is now in mL of gas produced.

5. To convert mL of gas into percentage of theoretical gas produced, divide the mL by the predicted observable total theoretical mL of gas from worksheet.

7.0 Documentation/Record Keeping

All records necessary to reconstruct the study and demonstrate adherence to established protocols will be maintained.

8.0 References

1. British Standard ISO 11734:1995, Section 5.21, Evaluation of the "Ultimate" Anaerobic Biodegrability of Organic Compounds in Digester Sludge—Method by measurement of biogas production.
2. OECD Guideline 311 (Proposal as of 10/01) for Ready Anaerobic Biodegradability—Gas Production from Diluted Anaerobic Sewage Sludge.
3. ASTM E1196-92, Determining the Anaerobic Biodegradation Potential of Organic Chemicals.

9.0 Attachments

Attachment 1—Test Apparatus

1. Assembly of Pressure Reading Devices
2. Preparation of Dispensing Station and Sparge Lines
3. Calibration and Leak Check
4. Values Derived from Calibration

Attachment 2—Worksheets

Anaerobic Inoculum Preparation Form
Preparation of Anaerobic Test Medium Form
Anaerobic Total Solids Measurement Form
Anaerobic Respirometer Test Design Sheet

ATTACHMENT 1

The Test Apparatus

1. Assembly of Pressure Reading Devices

32 - Stainless steel three-way valve with 1/16″ fittings (Swagelok, SS-44xS1)
32 - PX26 series (gauge) pressure transducer 0-15 PSIG, 100MV
32 - SS Reducing Union (Swagelok, SS300-6-1)
32 - Teflon Ferrules 3/16″, two piece (Swagelok, T-303-1, T-304-1)
64 - SS 16-gauge needle with Telfon Luer hub (Alltech KF-716)

Cut two of the 2″ SS needles into three pieces:

1. ½″ containing Teflon Luer Hub
2. 1 ¼″ containing the pointed end
3. ¾″ piece—straight SS tubing only

The valve is prepared by inserting tubing piece #2 into the base of the three-way valve (pointed end away from valve). The reducing union is connected to one side of the valve using tubing piece #3. Loosen the 3/16″ fitting on the reducer, insert the pressure transducer (PT) and tighten. If the PT will not tighten down, change out the two-piece Teflon ferrule on the nut. Piece #1 is inserted into the end of the valve opposite of the reducer fitting (hub facing outward).

2. Preparation of Dispensing Station and Sparge Lines

Omnispence liquid dispenser (Wheaton, cat # 375010/015)
3 mm silicon tubing (VWR 25485-007)
Fume Hood
Stir bars
One large (12 x12″ stir top dim) stir plate
3 L round bottom boiling flask
Heating mantle and controller
Pipetting needles, SS Hub, 18 gauge, 4″ (Popper & Sons)

NOTE: The area in a single fume hood is needed. Place the stir plate in between the dispenser and the heating mantle when setting up.

The Wheaton Omnispence is used to deliver the reduced media to the serum bottles. Cut ~6′ length of 3 mm diameter silicon tubing for the dispenser. The length of the tubing needs to reach from the bottom of the boiling flask (on stir plate), through the dispenser pump head, and have ~ 2′ of length remaining to be able to dispense the media into serum bottles in the fume hood. Attach the borosilicate glass sinker to one end of the tubing. Prepare the opposite end of the media line by attaching one of the 18 gauge needles bent as described in Figure A-2.

The nitrogen source is split (accessible in a fume hood) to allow sparging of the media and the dispensing line simultaneously. To create the sparge lines, hook up ~2′ of 3 mm silicon tubing to the N_2 outlet and clamp tubing in place. To split the line, add a three-way stop cock and connect two tubing extensions—one to reach the neck of the media flask (on stir plate) and the other to match the length of the media dispensing line. Insert a SS needle into the extension that reaches to the flask. Bend the needle as shown in Figure A-2b. To add a sparge line to the media dispensing line, insert an SS needle into the other silicon tubing extension and bend as shown above in Figure A-2a. Clamp the dispensing and sparge line needles together so that the sparge line is on the bottom as shown in Figure A-2a.

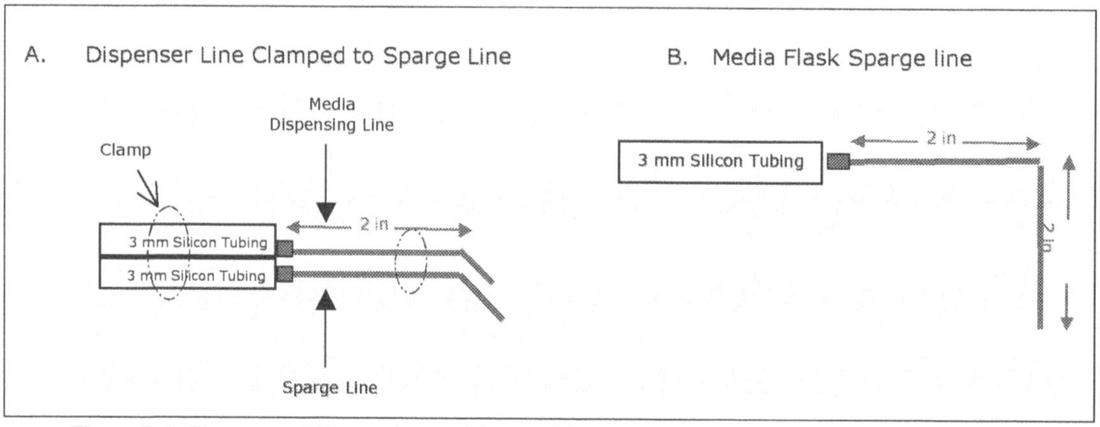

Figure A-2. Diagrams of Dispensing and Sparge Lines

3. Calibration and Leak Check

Source of CO_2 and CH_4 gas
Silicon Tubing
~1 L bottle with fitted stopper (stopper contains three holes)
Gastight syringe 20–50 ml with luer fitting
Luer lock needle
Silicone tubing

Connect silicon tubing to the end of the gas regulator fitting. Secure stopper on bottle. Carefully insert needle into one hole of stopper. Thread the silicon tubing from the gas source through a second hole of the stopper. The third hole serves as a vent. Turn on the gas to a gentle flow, enough so that there is positive pressure in the system. To calibrate the channels, a known volume of gas is injected into sealed serum bottles that contain 50 mL of test media with a gastight syringe. The volume of gas used is typically 5, 10, 20, 30, and 40 mL, and each gas volume is tested in triplicate. Set the sampling interval for 30 minutes. Once the system has reached equilibrium (at least 24 hours) the final reading is taken. Subtract the starting mv value ("zero value") from the final reading. Plot the volume of gas injected against the mv output of the instrument. In addition to generating calibration data, this test is used to check for leaks in the system. Observe the data trend over the 24-hour period for all channels. If the mv value shows a steady decline after the gas was added, there is likely a leak in the pressure transducer assembly. If this is the case, check for leaks at connections on the assembly and/or replace the Teflon ferrule in the nut that connects the pressure transducer to the reducer.

4. Values Derived from Calibration

The data from the calibration are used as input for converting mv into mL of gas produced. Perform linear regression analysis (Jandel Tablecurve v3 or other software) for each gas by plotting mv versus mL of gas.

$$\text{mL of } CO_2 = a + bx$$

$$\text{mL of } CH_4 = a + bx$$

where a = x intercept and b = slope

The percent of CO_2 predicted to remain in solution during respiration is determined and corrected for the course of the test.

$$\text{Slope } CO_2/\text{slope } CH_4 = \% \text{ of } CO_2 \text{ in headspace}$$

This percentage is used when completing the theoretical gas production worksheet for a material.

$$\text{Theoretical mL of } CO_2 * a + \text{Theoretical mL of } CH_4 = \text{Observable Total Theoretical Gas (mL)}$$

This observable theoretical mL of gas is used to calculate the percentage mineralized during the course of the test. That is, after the mv readings from the instrument are converted to mL and corrected for background, they are divided by this total to determine percentage mineralized*.

* At the end of the test, when the data reaches a plateau, a sample of the media is removed from each serum bottle and analyzed for Dissolved Inorganic Carbon. The amount (in mg C/L) and the final mv reading from the instrument are summed to determine total gas production in mg of C.

ATTACHMENT 2

Worksheets

Anaerobic Inoculum Preparation Form

Date ____ Initials: ________________

Study No: ____

Inoculum Collection:

WWTP and Digester Number ____

Date ____

Inoculum Preparation

The level of inorganic carbon (IC) added to the serum bottles should be <10 mg/L. It is recommended that DIC analysis be performed on the inoculum prior to the test set up and additional clean up be conducted if the IC level is too high.

Description for inoculum preparation:	Check or Init.
1. Fill two 250 mL centrifuge tubes with digester solids and spin at a low speed (e.g. 3000 rpm) for up to five minutes.	________
2. Carefully pour off the supernatant in a waste container while sparging the headspace of the centrifuge tube with nitrogen.	________
3. Re-suspend each pellet by filling the tube with approximately 200 mL of the media using the Omnispenses dispenser.	________
4. The "washed" solids are mixed and centrifuged (conditions above).	________
5. The supernatant is removed as previously described. The pellet is resuspended with approximately the same volume of fresh media as described above.	________
6. Remove two 10-mL subsamples for DIC analysis and seal in 15 mL centrifuge tubes. Transfer inoculum to 0.5 L amber bottle, add a stir bar, and seal with a gas lock stopper assembly.	________
7. Place bottle on stir plate, remove stopper and while sparging the headspace of the bottle, remove 3-5 replicate 1 mL samples for total solids (TS) analysis. Record analysis on solids sheet.	________
8. Fill the gas lock device with water. Incubate the bottle in the 35°C oven.	________
9. Centrifuge the inoculum samples on a table-top centrifuge until supernatant is clear. Directly analyze supernatant sample for DIC without dilution. (Target a range of 50-75 mg/L IC when adding 5-10 mL of inoculum to a 50 mL total test volume).	________
Note: If further washing of the solids is required to lower the IC level, document in the study record.	________

Preparation of Anaerobic Test Medium Form

Date ________________________________ Initials: ______________

Study No: ________________________________

Volume Prepared ________________________________

Reagent	Manufacturer/ Lot #	Amount added/1 L	Amount added/2.5 L	Amount Weighed
		(g)	(g)	(g)
KH_2PO_4		0.27	0.68	
Na2HPO4-12H_2O		1.12	2.8	
NH_4Cl		0.53	1.33	
$CaCl_2$-2H_2O		0.08	0.19	
$MgCl_2$-6H_2O		0.10	0.25	
$FeCl_2$-4H_2O		0.02	0.05	
*Resazurin		1 ml	2.5 ml	
Na_2S		0.03	0.08	
* Prepare 0.1g/100 ml stock solution				

Description below for preparation of 2.5 L of media: Check or Init.

1. Add 2.5 L of water, stir bar and boiling chips to the 3L flask. Place the flask in the heating mantle, place mantle on the large stir plate and turn on stirrer. Clamp a thermometer in place so that it will accurately measure the temperature of the solution. ________
2. Add the reagents listed above except for the Na_2S and continue stirring the media throughout the preparation. Place the media sparge line in the flask, ensuring the tip of the needle reaches below the base of the neck, and secure with a clamp attached to ring stand. Do not turn on N_2 at this time. Heat the mantle to 105°C until the solution has boiled for approximately five minutes. ________
3. Turn the dial down to 35°C and turn on the N_2 line to maintain the media in a reduced condition continuously sparged with N_2, or seal with a stopper. *Note: To cool the media faster, take it out of the heating mantle temporarily. Use the mantle at 35°C when dispensing the media.* ________
4. Add the reducing agent, Na_2S, after the media temperature falls below 45° C. The solution should turn from light pink (intermediate reduction to resorufin) to a clear solution (completely reduced) with black precipitate. The complete reduction may take several hours. If the media is not completely reduced by the time it is to be used, continue with the experiment because the addition of the inoculum should reduce the media within a 24-hour incubation period. If the reduction does not occur within this 24-hour timeframe, the test is aborted and restarted. ________

Anaerobic Total Solids Measurement Form

Date ____ Initials: ________________
Study No: ______________________________________

Procedure:

1. Use ProWeigh Filters DoubleWeigh Filters.
   Preweighed 47mm glass fiber filters for Total Suspended Solids.
   Environmental Express, 490 Wando Park Blvd., Mt. Pleasant, SC 29484
   1-800-343-5319
2. Dispense well mixed sample of anaerobic digester solids to filter.
3. Dry for at least one hour at 103°C to 110°C.
4. Cool in a desiccator and weigh.
5. Samples should be run in at least triplicate.

Balance Used ____
Oven Used: ____
Oven Temperature: ____

<u>Calculation: (if using 1 ml)</u>: g total residue/L = <u>(A-B) X 1000</u>

A = weight of filter + residue, mg
B = weight of filter, mg

Sample Description	mL of sample	Filter wt (g)	Filter + Sample (g)	Results (g/L)
			Mean	

Water Environment Research Foundation

Guide to level of solids needed based on Carbon level in Test

Carbon level in Test (mg/L)	Solids Level	Solids/50 ml
≥100 mg/L	3%	0.15g
≥ 100 mg/L	2%	0.10g
≥ 50 mg/L	1%	0.05g

Anaerobic Respirometer Test Design Sheet

Date ____ Initials: ______________

Study No: ________________________________

Channel No.	Compound	% Carbon	Target Substrate volume (mg or mL)	Measured Substrate volume (mg or mL)	Inocolum volume (mL)	Media volume (mL)
1						
2						
3						
4						
5						
6						
7						
8						
9						
10						
11						
12						
13						
14						
15						
16						

A.7 Physical Disintegration (Vortex) Test

1.0 Purpose

This test is used to determine the length of time before a test material will begin to break apart under the physical force of swirling water.

2.0 Scope

The vortex test is used to evaluate the physical disintegration potential of a test material. The test materials may include tissue, wipes or any other flushable product. This test is used to gain a better understanding of the material properties and is not used for assessing flushability.

3.0 Principles

A beaker filled with tap water is used to measure the physical disintegration of a test material. The water in the beaker is set in motion with a stir bar. The time at which the first piece of product is broken off is recorded.

4.0 Responsibilities

Researchers in the Global Flushability Lab or contract persons reporting to them are responsible for conducting this test.

5.0 Safe Practices

Follow normal safety precautions when handling any test materials or laboratory equipment.

6.0 Procedure

6.1 Experimental Design

A 2 L beaker is filled with room temperature tap water (to the 2 L mark) and then placed on a magnetic stir plate. A stir bar (octagonal shape 2″ long and 3/8″ high) is placed in the water. The speed of the stir bar is set at 650 rpm. Once the speed is set, the magnetic stir plate is switched off using the toggle switch. The test material is then added to the breaker and the timer is started. For bath tissue, one sheet is folded lightly into quarters without creasing prior to dropping into the beaker of water. After five seconds, the magnetic stirrer is switched on at the preset speed. The endpoint of the test is when the first piece of test material breaks off. The timer is stopped and the reading is recorded. If the test material knocks the stir bar out of its rotation, both the timer and the stir plate are switched off and then back on again to restart the stir bar. If the test material continues to interfere with the stir bar, the test is terminated.

6.2 Concentration or Dosage Levels

The vortex test is run with one piece of substrate, approximately 100mm × 100mm.

6.3 Sample Preparation

Tissues are carefully separated at the perforation or the substrate is carefully cut so that the sample to be tested is the size of a single sheet of bath tissue. There should be no ragged edges or excess tissue at the perforation. The square of substrate is lightly folded into quarters without creasing. Other products may be added to the beaker as they would appear during use.

6.4 Type and Frequency of Tests, Analyses and Measurements

Five replicates are run for each test material with the end point being the time it takes for the first piece to break off the substrate. If a replicate is terminated due to interference with the stir bar, extra replicates are run so that an average of five replicates can be obtained. If the sample doesn't begin to disintegrate within 10 minutes, the test is terminated.

6.5 Data Reporting

This test is to evaluate the disintegration characteristics of a substrate relative to similar type products. The test report will include the average (± std dev) disintegration time plus any observations or comments concerning the behavior of the test material in the test.

6.6 Statistical Methods

Statistical methods if used will be described in the study records.

7.0 Documentation/Record Keeping

All records necessary to reconstruct the study and demonstrate adherence to established protocols will be maintained.

8.0 References

None

9.0 Attachments

None

A.8 Hydrolysis/Solubility Test

1.0 Purpose

This test is used to assess the disintegration of a flushable product via hydrolysis and/or solubility in an aqueous environment.

2.0 Scope

Hydrolysis and/or solubility may influence the degradation of products once disposed down a toilet. This test method is used to evaluate the rate and extent of hydrolysis and/or solubility of a flushable product in an aqueous environment. Since this test is conducted in tap water, caution should be taken when extrapolating the test results to wastewater environments. Additional testing and data may be needed to fully understand these loss mechanisms.

3.0 Principles

This test method is similar to the aerobic disintegration shake-flask test methods, except that in this method tap water is used—as opposed to raw wastewater—and the flask is static—as opposed to being shaken. Thus, only product dissolution into water is considered. In this test, the whole product is placed in a 3 L flask containing 1 L of tap water. Periodically, flasks are removed and their content is passed through a series of sieves. The various size fraction retained on the sieves are weighed and the rate and extent of hydrolysis/solubility is determined.

4.0 Responsibilities

Researchers in the Global Flushability Lab or contract persons reporting to them are responsible for conducting this test.

5.0 Safe Practices

Follow normal safety precautions when handling any chemicals.

6.0 Procedure

6.1 Materials and Apparatus

The following materials are needed to perform this test.

1. Triple Baffled Fernbach Erlenmeyer Flasks (2800 mL), Bellco Glass Inc., Catalog number 2551-02800.
2. USA Standard Testing Sieves: 5/16" (8mm), #5 (4mm), #10 (2mm), #18 (1 mm).
3. Drying oven capable of maintaining a temperature of 40 ± 3°C for plastic test materials and capable of maintaining a temperature of 103 ± 3°C for non-plastic test materials.
4. Whatman #41 filter paper, 12.5 cm circles.
5. Shower head and hose assembly with rubber fitting for sink faucet.
6. Aluminum weighing dishes, two sizes: 43mm and 57mm.
7. Forceps.
8. Commercial paint brushes from hardware store.

6.2 Test Initiation

Each test product is run in triplicate to accommodate three predetermined destructive sampling time points. Each flask will contain 1 L of tap water and 3–5 g of the test product. Each test product should be pre-weighed in triplicate on a four-place balance and these weights recorded in the laboratory notebook for final percent disintegration calculations.

Measure 1 L of tap water into each of the Fernbach flasks and place them on the bench top. Add 3–5 g of test product to each flask. Visual observations of the test product and the control product should be recorded after one hour. The recommended sampling times for most products are at six hours, 24 hours and 48 hours. Depending on the rate of hydrolysis/solubility, shorter or longer time periods may be used.

6.3 Test Termination

At the designated destructive sampling points, remove a flask from each set of products being tested. Pour the entire content of a single flask through a nest of sieves arranged from top to bottom in the following order: 8 mm, 4 mm, 2 mm and 1 mm. With a handheld showerhead spray nozzle, gently rinse the material through the nested sieves being careful not to force passage of the retained material through the next smaller sieve. After several minutes of rinsing, remove the top sieve and continue to rinse the next smaller sieve, still nested, for a few minutes. Be careful not to force passage of retained material to the next smaller sieve. Continue this rinsing process until all of the sieves have been rinsed. After rinsing is complete, remove the retained material from each of the screens using forceps and/or commercial paintbrushes. Transfer the content from each screen to a separate, labeled aluminum pan. Place the pan in the oven overnight at 103 ± 3°C (or other appropriate temperature) until the sample is dry. Continue this procedure at each designated sampling time point until all the test products are sampled. After all the samples are dry, weigh the materials from each of the retained fractions and calculate the percentage of hydrolysis/solubility based on the initial starting weight of the test material.

6.4 Suggestions for Sampling

Begin removing material from the smallest sieve (1.0 mm) first. Remove the material retained on the smallest screen and transfer it to an aluminum pan. Take the next larger sieve (2.0 mm) and invert the screen on top of the smallest (1.0 mm) sieve. Rinse the retained material from the larger screen onto the smaller one, then collect the material from the smallest sieve and transfer it to an aluminum pan. Repeat this procedure for the remaining sieves, continuing to transfer the content of the larger sieves to the smallest (1.0 mm) sieve and ultimately to an aluminum pan for drying.

6.5 Calculation

Calculate the percent of loss of test product at each time point using the following equation:

$$\frac{\text{Dried weight of retained fraction on sieve (gm)}}{\text{Initial dried weight of sample (gm)}} \times 100 = \%\text{ of initial added mass}$$

7.0 Documentation/Record Keeping

All records necessary to reconstruct the study and demonstrate adherence to established protocols will be maintained.

8.0 References

ASTM E 895—89 (Reapproved 1995) Standard Practice for Determination of Hydrolysis Rate Constants of Organic Chemicals in Aqueous Solutions.

9.0 Attachments

None

A.9 Confirmatory Toilet Bowl and Drainline Clearance Test

1.0 Purpose

This test is used to determine the relative clogging potential of a product as it passes through a toilet and household drainage system.

2.0 Scope

This test method evaluates the relative ease with which a product passes through a toilet and household drainage system. The test uses actual toilets and drainlines that are set to specific country plumbing codes. A positive result in this test would indicate the test product is compatible with household toilets and plumbing systems.

3.0 Principles

The test system consists of toilets and drainlines that are representative of those found in North America, Europe, and Asia. Flushable products are evaluated for their relative ease of evacuating the toilet bowl and trap, as well as being transported through a simulated drainline plumbing system. This confirmatory evaluation typically simulates two days of normal toilet use by a family of four (based on habits and practices data). The flushing condition includes male urination (no product), female urination (bath tissue), bowel movement (bath tissue along with SFM), and any other product of interest. Test products are evaluated relative to currently marketed products with acceptable flushability characteristics.

4.0 Responsibilities

Researchers in the Global Flushability Lab or contract persons reporting to them are responsible for conducting this test.

5.0 Safe Practices

Immediately clean up any spilled water or simulated fecal material (SFM) from the test apparatus because it may cause a slipping hazard. Wear slip resistant shoes for stepping up on the test stand.

6.0 Procedure

6.1 Test System Design

The test system consists of two sets of toilets and drainlines. The first test stand has three toilets connected to a common 100 mm ID (4″) drainpipe The total length of the drainline, depending on the position of the toilet on the test stand, should be > 60′ (18 m). The drainline includes four bends constructed of two 45° elbows. The line pressure of the water supply to the toilets is 20 ± 3 PSI. A standard 4″ ID (100 mm) “T” is inserted at the junction between the toilet drainline and the test drainpipe. The second test stand has four toilets with each connected to one of two 75 mm ID (3″) drainlines. The total length of the 75 mm drainline should be > 60′ (18 m), depending on the position of the toilet on the test stand. The slope of each of the drainlines is adjustable to meet the building codes of different regions. The required slope will be noted in the protocol and study records.

6.2 Experimental Design

The confirmatory test procedure simulates two days of normal toilet usage for a family of four (two males, two females), based on habits and practices data. The flushing conditions simulate male urination visits (no product), female urination visits (with bath tissue), and bowel movement visits (SFM with bath tissue). The SFM is loaded at 160 ± 15 g per flush. This loading of SFM is based on the average amount of fecal matter disposed of per person per day. The product loading is the average amount of product an individual would use per day, based on habits and practices data. The sequence provides for a base size of 20 product flushes (14 product-only flushes and six product and simulated fecal material flushes) as well as 14 empty flushes simulating male urination per test. Triplicate test runs (total base size of 60 product/102 total flushes) are performed for each product per loading level.

6.3 General Flushing Instructions

1. At the start of the test, flush the toilet to be used to set the correct water level in the bowl and clear any residue that may be in the drainline.
2. Place each piece of SFM or product into the toilet at 10-second intervals to allow time for the product to wet out.
3. Flush the toilet and observe the evacuation of the bowl.
4. Watch for the SFM and products to enter the drainline.
5. Enter data for each flush on the Attachment 2 worksheet by following the tables in Attachment 3.

Unless otherwise indicated in the protocol or study records, the flush sequence will be as follows:

1. **Flush with tissue only**—Record the bowl/trap clearance data. Record the distance the product travels in the drainline for that flush. If the product clears the drainline, designate that in the correct column (see the worksheet sample in Attachment 2). If the products completely cleared the drainline, skip the next empty bowl flush and go to Step 3 once the toilet tank has refilled.

2. **Flush with empty bowl**—Record the distance the product(s) travel after this flush. Go to Step 3.
3. **Flush with tissue only**—Record the bowl/trap clearance data. Record the distance each product in the drainline travels. Go to step 4 unless the drainline has cleared, then go to step 5.
4. **Flush with empty bowl**—Record the distance the product(s) travel after this flush. Go to step 5.
5. **Flush with tissue and SFM**—Record the bowl/trap clearance data. Record the distance each product travels in the drainline. Go to step 6.
6. **Repeat previous sequence (steps 1-5) five additional times**—Step 6 consists of 25 total flushes. After completing step 6 move to step 7.
7. **Flush with tissue only**—Record the bowl/trap clearance data. Record the distance each product in the drainline travels. Go to step 8 unless the drainline has cleared, then go to step 9.
8. **Flush with empty bowl**—Record the distance the product(s) travel after this flush. Go to step 9.
9. **Flush with tissue only**—Record the bowl/trap clearance data. Record the distance each product travels in the drainline. Go to step 10 unless the drainline has cleared, then the flush sequence is complete.
10. **Flush with empty bowl**—Record the distance the product(s) travel after this flush. If all the solids have not cleared the drainline after this flush, continue the empty flushes until the drainline is clear, recording the distance traveled after each flush. The maximum number of extra flushes allowed to clear the drainline will be stated in the study records.
11. If at any time in the flushing sequence the product remains in the toilet bowl or trap after flushing, the toilet is either flushed again or the tissue is plunged into the drainline manually and the flushing sequence continues. After completion of each trial (approximately 34 flushes), the drainline will be cleared prior to beginning subsequent testing.

6.4 Concentration or Dosage Levels

The confirmatory test is typically run at the 50th percentile loading, based on P&G Habits and Practices data from each region. In the absence of H&P data, expected usage scenarios provided by Product Development can be used. The loading will be documented in the protocol or study records.

6.5 Sample Preparation

In the confirmatory test, each strip of bath tissue is folded to the size of one sheet of tissue (approximately 4″ × 4″) before placing in the toilet. Folding the bath tissue has been shown in previous tests to be more conservative (i.e., harder to flush) than crumpling the tissue. Other products will be placed in the toilet according to their expected usage.

6.6 Type and Frequency of Tests, Analyses and Measurements

The confirmatory flush test will be run in triplicate. The following types of data will be recorded on the worksheets as instructed on Attachment 3:

1. Bowl/trap clearance.
2. Distance the product travels on each flush.

6.7 Data Reporting

The confirmatory flushability data are compared to data for a battery of reference products representing the acceptable range of flushability for currently marketed products. Results will be documented in a study report.

The data from each set of replicates will be displayed on the following tables:

Bowl/Trap Clearance Data

Product	Clogged bowl (%)	Clogged trap (%)	Number of plunges needed to clear clogs	Average distance traveled per flush	Average number of flushes to clear drainline

6.8 Statistical Methods

Statistical methods, if used, will be described in the study records.

7.0 Documentation/Record Keeping

All records necessary to reconstruct the study and demonstrate adherence to established protocols will be maintained.

8.0 References

1. ANSI Standard A112.19.2M-1998
2. R. B. Martin, Plumbing Standards October 1989, page 17
3. ANSI A112.19.6-1995 Hydraulic Performance for Water Closets and Urinals
4. Japanese Industrial Standard A 5207 (1995)
5. Standard Sewerage Ordinance and Methods for the Examination of Flushability of Sanitary Products (Japan)

9.0 Attachments

Attachment 1—Preparation of SFM
Attachment 2—Worksheet
Attachment 3—Instructions for taking toilet and drainline measurements

ATTACHMENT 1

Preparation of Simulated Fecal Material

Purpose

Simulated Fecal Material (SFM) is used to simulate feces being flushed in a toilet.

Scope

Feclone is purchased from SiliClone Studios as a dry powder and must be mixed with water and placed in tubes to correctly mold the shape.

Responsibilities

Technician prepares and stores the Feclone and keeps track of the supply and notifies lab manager of the need to order more.

Schedule

The preparation of SFM is carried out on an as needed basis.

Safe Practices

1. Unplug the mixer and extruder while assembling or disassembling the equipment.
2. Always wear insulated gloves when carrying and pouring the boiling water.
3. Pour water slowly to avoid splashing.
4. Keep hands clear of mixer while it is running.
5. While loading SFM into the extruder, keep hands out of chute.
6. Immediately clean up any spills to prevent a slipping hazard.

Procedure

1. Heat 2900 ml of tap water to boiling in three 2 L flasks.
2. Slowly pour the water into the mixing bowl of the Hobart mixer.
3. Add one bag (approximately 900 g) of Feclone BFPS-7 dry concentrate.
4. Add 20 ml of 1.3% Dawn solution. (500 ml water and 6.5 ml of Ultra Dawn)
5. Mix on low for one minute.
6. Stop the mixer and scrape the sides of the bowl to mix in all the dry powder.

7. Increase the speed to medium and continue mixing an additional two minutes or until all dry powder is mixed in.
8. When the material is well-mixed, transfer to the extruder pan.
9. Cut small squares of SFM and feed them down the chute.
10. Extrude the SFM into 50 ml disposable centrifuge tubes that have a small hole punched in the end of each.
11. Place screw caps on the end of each tube and place in the refrigerator if SFM will be used within the next three days. If the storage time will be longer than three days, store in the freezer to prevent fungal growth.
12. Before using, place in the water bath at 38°C.
13. After using the SFM, the tubes should be washed in a soapy Dawn (or a glassware cleaner) solution and scrubbed with a brush to remove all remaining SFM.
14. Soak the tubes in a 10% Clorox solution, rinse and air dry before refilling.

ATTACHMENT 2

Study No. ______________________ Test System ______________ Date ________ Position #1
Test Material ______________________ Test Conditions ______________________
Trial # ______________________ SFM ______________________

				Location in Drainline				
Flush #	Bowl Loading	Remaining in Bowl	Remaining in Trap	0-4 m u,p,1s,1e,2s	5-8 m 2s,2e,3s	9-12.5 m 3s,3e,4s,4e,5s	Clear Drainline	Note #

Notes:

Water Environment Research Foundation

ATTACHMENT 3

Table A-4. Bowl and Trap Clearance

Condition	Record in column headed, "Remaining in bowl"	Record in column headed, "Remaining in trap"	Next step
The product hesitates in the bowl and water level rises before it all clears (labored flush).	LF	0	Go on to the next product flush. [1]
Bowl is empty after flush and all solids are in the drainline.	0	0	Go on to the next product flush. [1]
Bowl is empty after flush but not all solids are in the drainline. The water level is normal.	0	List whatever is presumed to be in the trap. *Example: T1 for tissue from the 1st flush.*	Go on to the next product flush. [1]
Bowl is empty after flush but not all solids are in the drainline. The water level is high.	0	List whatever is presumed to be in the trap. *Example: T1 for tissue from the 1st flush.*	Plunge the product into the drainline. [1,2]
Product is still in the bowl after the flush.	List whatever is remaining in the bowl.	If all the solids can be accounted for between the bowl and drainline–0. Otherwise, list what is presumed to be in the trap.	Try an extra flush. If that doesn't clear the bowl, use the plunger to push the product into the drainline. [1,2]
Product is stuck in the trap, visible in the bowl but no product is floating in the bowl.	0	List whatever is presumed to be in the trap. *Example: T1 for tissue from the 1st flush.*	Use the plunger to push the product into the drainline. [1,2]

1. Wait until the products have stopped moving and the water is only trickling out of the drainline to start the next product loading. Record the location of the solids in the drainline (see Table A-5).
2. When plunging, one plunging incident consists of three quick plunges in succession. Record each plunging incident in the notes section. If the water level is low after plunging, reset with an extra flush.
3. Each extra flush is recorded on a new line along with the location of the solids in the drainline.

Table A-5. Drainline Clearance (Refer to worksheet on the following page)

Flush #	Explanation of Drainline Data	Next Step
1	The SFM and tissue cleared the toilet and ended up in the 1st straight pipe. The tissue traveled at least 2m but less than 3m	After the product has stopped moving in the drainline and the water slows to a trickle, start next loading.
2	S1T1 moved to 11m, the SFM from the 2nd flush (S2) stopped in 2E and the tissue (T2) stopped at 3m. S1T1 traveled to 11m.	After the product has stopped moving in the drainline and the water slows to a trickle, start next loading.
3	S3 cleared the toilet and stopped in 1E but all of the T3 tissue stayed in the trap. S2T2 traveled to 10m and S1T1 cleared the drainline.	Since the water level is normal, it may clear with just an extra flush.
4	The empty flush cleared the T3 tissue from the toilet and it traveled to 12m. S2T2 and S3 cleared the drainline.	Continue on with next loading as above.
5	The S4T4 loading hesitated before leaving the toilet but cleared without intervention. S4T4 went to 5m and T3 cleared the drainline.	Continue on with next loading as above.
6	The S5 cleared the toilet and was in the 1st elbow. Only one strip of tissue was in the bowl and none was visible in the drainline, so three strips of T5 must be in the trap. S4T4 traveled to 11m.	Since the water level is normal, it may clear with just an extra flush.
7	The extra flush failed to clear the toilet but S5 moved to 4E and S4T4 cleared the drainline. TheT5 tissue was plunged into the drainline and it traveled to the 1st elbow (4m).	Perform empty flush to reset water level in bowl after plunging and also to move the tissue through the drainline.
8	The empty flush reset the water level land moved the T5 loading.	Continue empty flush to clear the drainline.
9	T5 clears the drainline after empty flush.	Trial is complete.

Study No. PEL02-XXX Test System 100mm /UK Date 07/17/02 Position #1

Test Material LDT CLXXX Test Conditions 1.25% slope, no occlusion

Trial # SFM + 4x7 Charmin SFM 160g

				Location in Drainline				
Flush #	**Bowl Loading**	**Remaining in Bowl**	**Remaining in Trap**	**0-4 m u,p,1s,1e,2s**	**5-8 M 2s,2e,3s**	**9-12.5 m 3s,3e,4s,4e,5s**	**Clear Drainline**	**Note #**
1	S1T1	0	0	S1T1 (2m)				
2	S2T2	0	0	T2 (3m)	S2	S1T1 (11m)		
3	S3T3	0	T3	S3		S2T2(10m)	S1T1	
4	Empty flush	—	0			T3 (12m)	S2T2S3	
5	S4T4	LF	0		S4T4(5m)		T3	1
6	S5T5	T5d	T5abc	S5		S4T4(11m)		2
7	Extra Flush	T5d	T5abc			S5	S4T4	3
8	Empty flush	—	—			T5 (9m)	S5	4
9	Empty flush	—	—				T5	5

This worksheet is for a toilet in position #1 on the first test stand, which is connected to the 100 mm drainline. Use the appropriate worksheet corresponding to the position of the toilet and drainline in use.

m = meter if the location is between 2 meter markings, record as the lower number. Distance is only recorded for test product.

u = under the test stand

p = in the port (inverted T that is used as a drainline vent.)

s = straight pipe (the first straight pipe, between the port and the 1st elbow is designated 1s)

e = elbow (bend in the pipe constructed with 2-45° elbows).

Notes:
1. Labored flush—tissue hesitated and water level went up before clearing.
2. T5 consists of four strips of tissue. Three strips are in the trap and one strip remains in the bowl after flushing.
3. Empty flush failed to clear the toilet. 1XP (one plunging incident). After plunging, T5 was all in 1E (4m).
4. Empty flush to reset water level in the toilet bowl.
5. Empty flush to move the final loading through the drainline.

A.10 Tipping Pipe Disintegration Test (Proposed)

1.0 Purpose

This test is used to assess the fate of flushable consumer products during wastewater conveyance.

2.0 Scope

This test evaluates the fate and transport of a flushable product in sewer collection systems. Understanding a product's fate during conveyance is an important component when assessing the compatibility with municipal wastewater treatment systems. A positive result in this test would indicate that the product would be indistinguishable from other wastewater solids when it enters a municipal wastewater treatment plant. Moreover, this test method is being used to validate a computer mathematical model that predicts the fate and transport in flushable products in wastewater conveyance systems.

3.0 Principles

This test method is used as a secondary tiered bench-scale testing tool to evaluate the fate of flushable consumer products during wastewater conveyance. A tipping pipe apparatus is used in this assessment. In this test, a clear plastic pipe is loaded with product and tap water and the pipe is tilted up and down to simulate the flow within a sewer pipe. At specified time intervals, the apparatus is stopped and its content is passed through a series of sieves. The rate and extent of disintegration is then determined by measuring the weight of product retained on the various size fractions.

4.0 Responsibilities

Researchers in the Global Flushability Lab or contract persons reporting to them will be responsible for conducting this test.

5.0 Safe Practices

Follow normal safety precautions when handling any chemicals or laboratory equipment.

6.0 Procedure

6.1 Test Apparatus

The test apparatus consists of a 20′ long, 6″ diameter pipe mounted to a tipping stand. The pipe can be tilted ± 3% to cause fluid motion that mimics that of a sewer pipe. Through computational fluid hydraulic modeling, a 3% slope was determined to correspond with a maximum fluid velocity of 2′ per second in the pipe. The rate of ascent and descent of the pipe is controlled pneumatically via an electronic timer.

6.2 Test Product Preparation

Dry weight measurements of the test material should be made in triplicate prior to test initiation.

6.3 Testing Parameters

Tap water (or raw wastewater) is loaded into the pipe at 1/3 capacity. With a 20′ long and 6" diameter pipe, the volume of water to be added to the pipe is 37 L.

Slope is set at 3% and measured periodically throughout the test. This slope corresponds with a maximum velocity of 2′ per second.

6.4 Test Initiation

Pump 37 L of tap water into the each pipe using a peristaltic pump. Add approximately 7 g of test product to each pipe. For continuous cycle mode, set the desired cycle time on the control panel. Reset the batch set point to zero. Set the desired degree of tilting. Ensure there are no obstructions in the pipe's path. Push the start cycle. The green light will illuminate and the timer will start. The cycles will continue until the "Stop" button is pressed. In batch mode, the "Batch" button can be selected to set a number of cycles that are desired. The apparatus automatically stops when the number of cycles has been reached. The run time of the test will depend on the product being tested. An estimate can be made from the aerobic disintegration shake flask test results.

6.5 Test Termination

At the designated sampling points, the content of each pipe is collected in a 40 L basin. This sample is then poured through a nest of sieves arranged from top to bottom in the following order: 8 mm, 4 mm, 2 mm and 1 mm. With a handheld showerhead spray nozzle, gently rinse the material through the nested sieves, being careful not to force passage of the retained material through the next smaller sieve. After several minutes of rinsing, remove the top sieve and continue to rinse the next smaller sieve, still nested, for a few minutes, again being careful not to force passage of retained material to the next smaller sieve. Continue this rinsing process until all of the sieves have been rinsed. After rinsing is complete, remove the retained material from each of the screens by inverting the screen onto a smaller screen and rinsing the retained material in the reverse direction. This procedure is described in detail below. Transfer the contents from each screen to a separate, labeled aluminum pan. Place the pan in the oven (100 ± 3°C or other specified temperature) overnight or until the sample is dry. Continue this procedure at each designated sampling time point until all of the test products are sampled. After all the samples are dry, weigh each of the size fractions and calculate the percentage of disintegration based on the initial starting weight of the tested material.

6.6 Sieve Sampling

Begin by removing material from the smallest sieve (1.0 mm) first. Remove the material retained on the smallest screen and transfer to an aluminum pan. Then take the next larger sieve (2.0 mm) and invert the screen on top of the smallest (1.0 mm) sieve. Rinse the retained material from the larger screen onto the smaller one, and then collect the material from the smallest sieve and transfer to an aluminum pan. Repeat this procedure for the remaining sieves, continuing to transfer the content to the smallest (1.0 mm) sieve and ultimately to an aluminum pan for drying.

In products that contain a lot of pulp (i.e. bath tissue), it is recommended to pour a small portion of the content at a time through the nesting of sieves and rinse the material for a few minutes before adding the next portion. This allows the pulp to better separate on the sieves (especially on the smaller grid sieves) and not be susceptible to "tangling", which adds to unnecessary retention of material on the sieves.

6.7 Calculation

Calculate the percent of disintegrated test product using the following equation:

$$\frac{\text{Dried weight of retained fraction on sieve (gm)}}{\text{Initial weight of sample (gm)}} \times 100 = \%\text{ of initial added mass}$$

7.0 Documentation/Record Keeping

All records necessary to reconstruct the study and demonstrate adherence to established protocols will be maintained.

8.0 References

None

9.0 Attachments

Attachment 1—Tipping Pipe Apparatus

ATTACHMENT 1: TIPPING PIPE APPARATUS

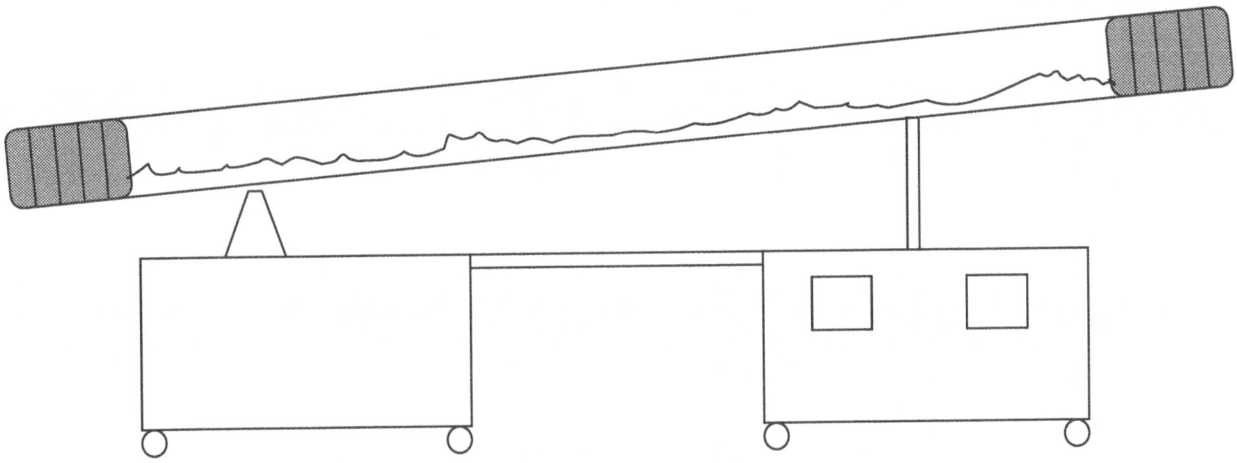

A.11 Porous Pot Disintegration Test

1.0 Purpose

This test is used to assess the fate of flushable products in aerobic biological treatment systems.

2.0 Scope

This test evaluates the potential for biologically mediated disintegration of flushable products in onsite aerobic and activated sludge treatment systems. A positive result in this test indicates that the flushable product will degrade in biological treatment systems. While this test does not assess biodegradation *per se* because it does measure gas production, it can give a good indication of whether the product is biodegradable.

3.0 Principles

This test method is used as a secondary tiered testing tool for assessing the fate of flushable products in onsite aerobic and activated sludge treatment systems. The test is conducted using a porous pot activated sludge test system. The system is seeded with activated sludge and dosed with raw wastewater collected from a local wastewater treatment plant. The test materials are placed in the porous pot reactor in a fiberglass mesh bag and their rate of disintegration is monitored over time.

4.0 Responsibilities

Researchers in the Environmental Fate Lab are responsible for conducting this test.

5.0 Safe Practices

This procedure involves the use of raw wastewater and activated sludge from a municipal wastewater treatment plant. Consequently, individuals performing this test may be exposed to microbiological agents that are potentially hazardous. Follow normal safety precautions when handling biologically active samples or any chemicals.

6.0 Procedure

6.1 Activated Sludge Collection

Activated sludge (AS) is collected from a local wastewater treatment plant and kept aerated until ready for use. Total suspended solids (TSS) of the AS is measured using Standard Methods and its concentration adjusted to a range from 2500 to 4500 mg/L by either concentrating or diluting the sludge. When concentrating, allow the AS to settle for 30 minutes and pour off a portion of the supernatant. When diluting, add a proportioned amount of raw wastewater to the AS.

6.2 Wastewater Collection

Raw wastewater is typically collected from the same wastewater treatment facility as the AS. Filter the wastewater through approximately 6 layers of cheesecloth to remove larger solids. This creates a nutrient-rich solution to be fed to the porous pot reactor. Store the filtered wastewater in a constant temperature room at 4 ± 2 °C with gentle mixing to keep the wastewater aerated and any remaining solids distributed.

6.3 Test Material Preparation

Weigh the test material on a 4-place analytical balance in triplicate and record its weight prior to placing in the porous pot reactor. Make a small mesh bag out of fiberglass screen (1.6 mm size opening) by stapling the seams. Encapsulate the material in the mesh bag. Attach a fishing weight to the bottom of the bag to help prevent it from floating in the aerated activated sludge. On the top side of the bag, attach either a thin wire or fishing line to provide a way to retrieve the mesh bags at designated sampling intervals.

6.4 Porous Pot Units

Set up the porous pot units within 48 hours of obtaining the activated sludge. Measure 6 L of well-mixed activated sludge into the porous pot, making sure the solids remain inside the stainless steel pot (or center membrane). The level of the liquor should be just below the effluent port. Place a stir bar in the pot and stir the sludge gently so that it remains well mixed at all times. Add an air line(s) to the reactor and adjust the flow so the activate sludge bubbles gently. The DO should be monitored and maintained within a range of 1–3 mg/L.

6.5 Test Initiation

Place the mesh bags containing test product into the center membrane of the porous pot reactor. Make sure the fishing weights are facing downward. Secure the top end of the bags to a bar suspended over the reactor or on the side of the glass jacket. The bags should be easily retrievable from the reactor.

6.6 Test Termination

At designated sampling time points, a mesh bag containing the test material is removed from the porous pot reactor and rinsed gently with tap water. Next, remove the test material from the mesh bag and place it on a 1 mm USA Standard Testing Sieve for further rinsing. Following rinsing with tap water, transfer the retained material on the sieve to a labeled aluminum pan. Place the pan in the oven at 103 ± 3°C (or other specified temperature) until dry.

After drying, weigh the remaining test material and calculate the percentage of disintegration based on the difference between thc initial weight and the final weight of the test material.

6.7 Calculation

Calculate the percent of disintegrated test product using the following equation:

$$\frac{\text{Final dried weight of test material (gm)}}{\text{Initial dried weight of test material (gm)}} \times 100 = \%\ \text{loss of mass}$$

7.0 Documentation/Record Keeping

All records necessary to reconstruct the study and demonstrate adherence to established protocols will be maintained.

8.0 References

None

9.0 Attachments

Attachment I—Porous Pot Diagram

ATTACHMENT 1

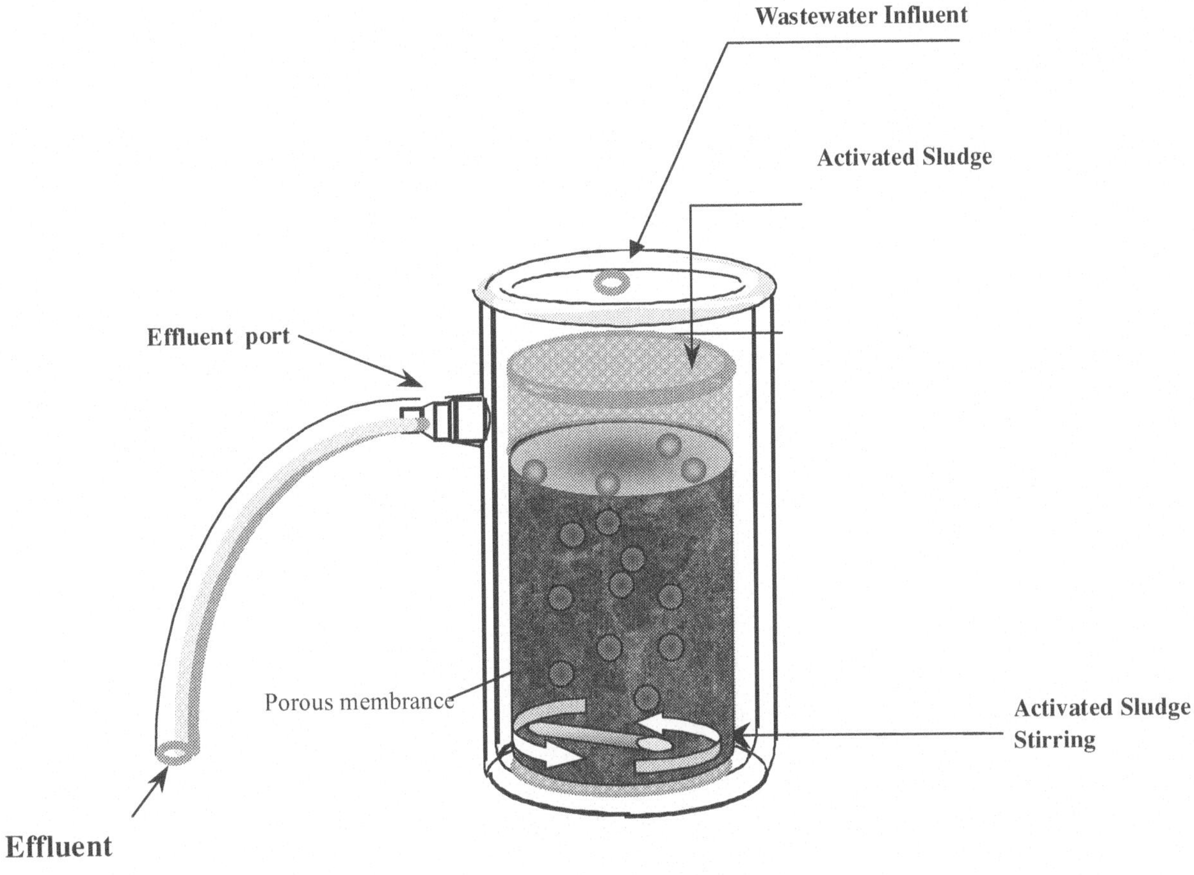

A.12 Soil Disintegration Test

1.0 Purpose

This test is used to determine the rate and extent of disintegration of a test material in a soil environment.

2.0 Scope

This test evaluates the fate of a flushable product following biosolids or septage amendment to soil. There may be certain instances in which a product may not biodegrade in the anaerobic environment of a septic tank or in an anaerobic digester, but it would do so in an aerobic soil environment. To be acceptable as flushable any product or product residue contained in the sludge should continue to disintegrate or degrade in the soil environment.

3.0 Principles

In this test, the material is either placed in a mesh bag (fiberglass screening with a 1.6 mm opening) and buried in soil or placed directly on the soil surface. Sludge or septage may also be amended at the same time. Periodically, a sample is retrieved to assess its disintegration. The content of the sample is placed on a 1 mm sieve and then rinsed with tap water. The material that is retained on the sieve is then dried and weighed. The weight loss of the test material over time is recorded.

4.0 Responsibilities

Researchers in the Global Flushability Lab or contract persons reporting to them are responsible for conducting this test.

5.0 Safe Practices

This procedure may involve the use of digester sludge from a municipal sewage treatment plant or septage from a septic tank. Consequently, individuals performing this test may be exposed to microbiological agents that are potentially hazardous. Follow normal safety precautions when handling biologically active samples or any chemicals.

6.0 Procedure

6.1 Experimental Design

Test materials are placed in mesh bags (fiberglass screening with a 1.6 mm size opening) and then buried in soil or placed directly on the soil surface. Sludge or septage may also be amended at the same time. Amendment rates should be typical of those practices in the field and documented in the study records. Samples are retrieved from the soil at designated time points. The content of these samples are then placed on a 1 mm sieve and rinsed with tap water. The material retained on the sieve is then dried (typically at 100 ± 3°C, but will depend on the melting point of the test material) and weighed. The percent of weight loss is calculated from the dried weight of the samples. Testing is limited to seasons when the ground is not frozen.

6.2 Concentration or Dosage Levels

Typically 3–5 g of test material is placed in each of the fiberglass mesh bag. A test usually consists of eight mesh bags for each test material. One bag of each product is to be harvested at each sampling time point.

6.3 Sample Preparation

Samples are dried and weighed before placing in the mesh bags. Other sample preparations will be listed in the protocol or study record. When samples are harvested, the content in the mesh bag is placed on a 1 mm sieve and then rinsed with tap water to remove any soil adhered to the test material. The test material that is retained on the sieve is then dried overnight before being weighed.

6.4 Soil Preparation and Maintenance

This test can either be conducted in the laboratory (proposed) or in the field. The laboratory test is conducted in a container with a surface area of approximately 2′ x 2′ and a depth of 6". On a weekly basis, 1″ of tap water is added to the surface of the container to simulate rainfall. The field test is conducted in a field plot with a surface area of 6′ x 8′ and a depth of approximately 1′. During the test period, temperature and weekly precipitation are recorded. If the rainfall does not exceed 1″ (2.5 cm) during the week, then tap water is added to the field plot to make up the difference. If there is no rain during the week approximately 30 gallons of water are added to the plot. The test plot is kept weeded during the testing period. Both test systems utilize a mixture of topsoil and peat moss. Septage or sludge (biosolids) could also be added. The ratios are set so that the final soil falls within a typical range of organic carbon content, cation exchange capacity and pH for an agricultural soil.

6.5 Type and Frequency of Tests, Analyses and Measurements

The sampling frequency will depend on the product being tested and will be documented in the study records. The data recorded for each sample are the dried weight, condition of the sample and any other relevant observations. Photos may also be taken of each sample. Other tests, analyses, tests and measurements will be included in the study record.

6.6 Data Reporting

The summary report will include the percent disintegration determined by weight loss, physical characteristics of the samples and photos if available.

6.7 Statistical Methods

Statistical methods, if used, will be described in the study records.

7.0 Documentation/Record Keeping

All records necessary to reconstruct the study and demonstrate adherence to established protocols will be maintained.

8.0 References

None

9.0 Attachments

None

A.13 Sewage Ejector And Grinder Pump Test

1.0 Purpose

This test is used to determine the compatibility of flushable products in wastewater pumping systems.

2.0 Scope

Residential plumbing systems typically rely on the force of gravity to provide discharge of their wastewater. Fixtures that are installed below the sewer lines require a means of transporting wastewater to the level of the main drainline. Sewage ejector pumps are more commonly used in these situations and have the ability to pump a high volume of water. These systems can also handle solids up to 2″ in size.

Gravity sewers are no longer the standard wastewater conveyance system. In neighborhoods where streets and sidewalks currently exist, the conversion from onsite treatment to centralized treatment is made with vacuum or pressure sewer systems.

In pressure sewer systems, a small pump is used to discharge to the main. There are two kinds of systems to provide the sewer pressure. Systems that use a septic tank effluent pump are referred to as STEP systems. STEP pressure sewers depend on septic tanks to first settle out solids, deliver wastewater to the narrow sewer piping and provide the necessary pressure to move it through the system. The other type of pressure sewer uses a grinder pump (GP). Wastewater from each property goes to a tank containing a pump with grinder blades. The pump grinds the wastewater solids to particles small enough to pass through the pump, valves and piping without clogging.

3.0 Principles

This test evaluates the compatibility of flushable product with sewage ejector and grinder pumps. In this test, a flushable product is dosed in 6 L of tap water to a catch basin where either an ejector or grinder pump resides. The pump is periodically activated, depending on the water-level sensor, and a portion of the contents is evacuated from the basin. The evacuated content is then pumped to a storage basin that is located approximately 8′ (2.4 m) above the pump basin. This elevation provides a pressure head for the pumps to lift against. The content in the upper basin is then screened prior to being reused in the lower basin. The system is loaded at a rate equivalent to one week of product use (based on habits and practices data) or until the system fails, whichever comes first. Pump failure occurs when the pump can no longer evacuate the product from the basin.

4.0 Responsibilities

Researchers in the Global Flushability Lab or contract persons reporting to them are responsible for conducting this test.

5.0 Safe Practices

Follow normal safety precautions when handling pumps or product.

6.0 Procedure

6.1 Materials and Apparatus

The main apparatus for the test is a pallet-rack test stand, which measures 10′ in height, 3′ in depth and 7′ in width, with a shelf 6-7′ above the ground. Other materials needed for the test are listed below.

1. Open-top, flat-bottom HDPE cylindrical tanks with a capacity of > 85 gallons.
2. Sewage ejector pumps (representative brands).
3. Sewage grinder pumps (representative brands and types).
4. Drainage tubing approximately 4″ in diameter.
5. Plumbing fittings, check values and hoses compatible with the selected pump systems.
6. 110-volt and 220-volt electrical connectivity.
7. Screens for collecting solids from the water.
8. Plumbing valves for installation onto the upper basins for drainage.

6.2 Test System Preparation

The pallet rack is assembled and anchored to the floor for additional support. Two HDPE basins are placed on the shelf of the pallet rack. Two additional basins are placed under the shelf, directly below the upper basins. A selected sewage pump is placed inside the basin. Piping is connected to the pump for evacuation delivery to the upper basins.

6.3 Test Initiation

The lower catch basin with pump is typically dosed with 6 L of tap water to simulate a pre-determined water closet volume, along with the test product. This loading of water and test product is repeated until the sensor of the pump is triggered, and the contents of the lower basin are pumped into the upper basin. The product loading is set to simulate the typical consumer toilet habits and practices for a period of one week or until the system fails, whichever comes first.

6.4 Collection of Data/Measurements

The following measurements and observations should be made during the test run.

1. Number of loadings to trigger sensor.
2. Volume of water loading.
3. Amount of product loading.
4. Observation/evaluation of product in basin.
5. Observation/evaluation of pump clogging.
6. Amount of product in basin at end of test run.
7. Product settleability.

7.0 Documentation/Record Keeping

All records necessary to reconstruct the study and demonstrate adherence to established protocols will be maintained.

8.0 References

None

9.0 Attachments

None

A.14 Home Use Toilet And Drainline Test

1.0 Purpose

This test is used to assess the compatibility of a flushable product with household plumbing fixtures under normal usage conditions.

2.0 Scope

This test evaluates the compatibility of a flushable product in an actual home setting. Product is provided to several homes for use. The household members are asked to keep a diary of any problems they experience while using the product. The results are compared to those of flushable products currently on the market.

3.0 Principles

In this test, the flushable product is provided to several households (typically 50–100) and the household members are asked to use the product as they normally would. The households are selected based upon their responses to a pretest questionnaire. Criteria for participation typically include family size, type of toilet, type of drainline, type of treatment system, and questions around product use. The in-life portion of the test typically lasts from one to two months, but could run longer depending on the product being tested. The household members are asked to keep a diary during the testing period. At the end of the study period, the household is asked to fill out a posttest questionnaire. The results are tabulated in terms of the number of incidents per person per month.

4.0 Responsibilities

This test is typically conducted by the Products Research group.

5.0 Safe Practices

Follow normal safety precautions when handling any products or chemicals.

6.0 Procedure

6.1 Test Design

The ultimate design of this test will depend on the product being evaluated, but in general, 50–100 households are recruited to participate. The households are selected based upon their responses to a pretest screening questionnaire. Criteria for participation typically include family size (three or more in household), type of toilet (age), type of drainage pipe (plastic, cast iron, etc.), type of treatment system (onsite or municipal), and product use information. An example of the criteria to participate as well as a pretest questionnaire are provided below.

Panelist Criteria

	Criteria	% Desired
Availability	At Home till mid July	100%
Family size	Three or more in the household One or two children between 3-10 yrs old	100% 20%
How long live there	> One year	100%
Sewer system	No septic system Septic system	95-90% 5–10%
Age of toilet	a) Homes built before 1995 b) Homes built after 1995	30% 70%

Pre-test Questionnaire

1. Demographics (age+sex+income)
2. How many people live in your household?
3. How many women, men and children in the family?
4. Age of each member of the family?
5. Approximately how long have you been living there?
6. Availability?
7. What type of toilet do you have at home?
8. For your toilet type, what is your flush system?
9. What type of sewage system do you have?
10. In the past six months, what are all the brands of toilet tissue that have been purchased for your household?
11. In the past six months, what *one* brand of toilet tissue has been purchased *most often* for your household?
12. House classification
13. When was the house constructed?
14. When was the last time your toilet has been remodeled?
15. When was the last time your pipes have been remodeled?
16. How many toilets do you have in your household?
17. What type of drainage pipe do you have at home?

18. What is the diameter of your drainage pipe?
19. Thinking just of flushing your toilets, please indicate the number of times that you have experienced the following problems for any of your toilets at home in the timeframes shown below.

The in-life portion of the test typically lasts one to two months. The household is asked to keep a diary during the test period, which includes how much product is being used and any problems that have occurred. Below is an example of an incidence questionnaire.

Incidence Questionnaire

1. Panelist name
2. Date
3. Time
4. Who performed the task?
5. Which of these problems did you observe?
6. How did you solve the problem?
7. How many sheets did you use when the incident occurred?
8. How did you prepare the sheets when you went to the bathroom?
9. Have you experienced a similar problem in the last 6-12 months?
10. If yes, do you recall which toilet tissue brand it was?

Following the in-life portion of the study the household is asked to complete a post-test questionnaire. An example of a post-test questionnaire is below.

Post-test Questionnaire 1st Month

1. How many rolls are left in your case at the end of this month's test?
2. Do you think you used about the same/more/less quantity of test product versus that of your usual brand?
3. When rinsing the bowl, did the removal of the solids/paper by the water happen comparably fast and as complete as with your usual brand?

Post-test Questionnaire 2nd Month

1. How many rolls are left in your case at the end of this month's test?
2. Do you think you used about the same/more/less quantity of test product versus that of your usual brand?
3. When rinsing the bowl, did the removal of the solids/paper by the water happen comparably fast and as complete as with your usual brand?

6.2 Data Reporting

The results from this study include the number of incidents per person per month or the number of incidents per product use per month.

6.3 Statistical Methods

Statistical methods if used will be described in the study records.

7.0 Documentation/Record Keeping

All records necessary to reconstruct the study and demonstrate adherence to established protocols will be maintained.

8.0 References

None

9.0 Attachments

None

A.15 Onsite Treatment Compatibility Test

1.0 Purpose

This test is used to determine the compatibility of a flushable product in a septic tank or aerobic onsite treatment system.

2.0 Scope

This test evaluates the compatibility of a flushable product with onsite wastewater treatment systems. In particular, this test assesses whether a product will be retained in a septic tank or aerobic systems and if there will be no harm to the operation of these systems. Flushable products that are retained in these systems and cause no operational problems are considered compatible.

3.0 Principles

This test is conducted in residential onsite wastewater treatment systems (septic tanks or aerobic systems). These full-scale field test systems have been slightly modified for dosing, observing and sampling. The product can either be dosed through a sample port or flushed down a toilet in the home. These onsite systems are monitored over time to determine if the product is retained in the tank and has settled with the other wastewater solids. In particular, observations are made for floating product through access ports and for product escaping the tank in an effluent trap. This home use test provides a realistic look at product compatibility with the onsite systems.

4.0 Responsibilities

Researchers in the Global Flushability Lab or contract persons reporting to them will be responsible for conducting this test.

5.0 Safe Practices

This procedure involves contact with septage from onsite wastewater treatment systems. Consequently, individuals performing this test may be exposed to microbiological agents that are potentially hazardous. Follow normal safety precautions when handling biologically active samples or any chemicals.

6.0 Procedure

6.1 Experimental Design

The onsite treatment systems in this test are used to assess the settling characteristics and compatibility of a flushable product. These systems (one septic system and one aerobic system) have been modified with portals to allow for easy access to the systems. These access portals allow for monitoring of the system to note whether the product settles or floats. In addition, an effluent trap has been installed in both systems (i.e., on the effluent pipe between the septic tank and the drainage field and on the effluent pipe between the aerobic chamber and the sand filter) to catch any products that are not retained in these systems. These traps serve a dual purpose—first, to protect the drainage field or sand filter and second, to alert us to the fact that a product has not settled properly.

In this test, a flushable product is dosed to the onsite treatment systems either through a dosing port or by flushing directly in the home. The dosing sequence is based on habits and practices data. The length of time a study runs will depend on the product, but a minimum of 6 months is required to assess full compatibility with these treatment systems. After completion of the study these systems may be pumped to assess the degradability of the product. The pumped septage is passed through a basket screen (1″ opening) and a catch screen (1/8″ opening) to assess whether the remaining product would reach the soil environment in a recognizable form during land application. The product retained on the screens will be collected, dried and weighed to assess its loss over time. Pictures will also be taken to document the products fate. Material collected on the catch screen may also be used in a soil disintegration study to support compatibility with this disposal practice.

6.2 Concentration or Dosage Levels

The amount of product and the method of dosing may vary depending on the purpose of the test, type of product and the expected usage patterns in the household. The product is either flushed in the home toilet on a scheduled basis in a non-use situation or flushed by members of the household as used. In the non-use situation, the amount of product being dosed is based on habits and practices data. Samples may also be dosed through a clean out port in the drainline between the house and the onsite treatment system. The method of dosing and the dosing level will be included in the protocol or study records.

6.3 Sample Preparation

Sample preparation, if any, will vary depending on the product and the method of dosing. Any sample preparation needed will be outlined in the protocol or study records.

6.4 Type and Frequency of Tests, Analyses and Measurements

The onsite treatment systems will be monitored on a regular basis during the study as indicated in the study records. Any additional analyses of product, septage, sludge layer, scum layer or effluent will be recorded in the study records.

6.5 Data Reporting

The data collected will be used to determine if the product can be successfully disposed to an onsite wastewater treatment system. A summary report will be prepared at the conclusion of the study.

6.6 Statistical Methods

Statistical methods, if used, will be described in the study records.

7.0 Documentation/Record Keeping

All records necessary to reconstruct the study and demonstrate adherence to established protocols will be maintained.

8.0 References

None

9.0 Attachments

None

A.16 Onsite Treatment Degradation Test

1.0 Purpose

This test is used to determine the rate and extent of degradation of a test material in a septic tank or aerobic onsite treatment system.

2.0 Scope

This test evaluates the degradability of a test material in onsite treatment systems. A rapid disintegration in this test indicates that the flushable product will degrade in these treatment systems in a reasonable period of time.

3.0 Principles

The home use test provides a realistic look at product compatibility with the onsite systems. This testing is conducted in residential onsite systems (septic tanks or aerobic systems) that have been slightly modified for dosing, observing and sampling. Pre-weighed products in mesh bags (fiberglass screening with a 1.6 mm size opening) are suspended in the tanks and then recovered at designated time intervals to measure the rate of degradation by weight loss. If a flushable product does not begin to degrade in a septic tank or aerobic onsite system, it may increase the depth of the sludge layer and result in the need to pump out the system more frequently. Products that remain intact in the onsite treatment systems will be more visible in the environment during and after land application of the septage.

4.0 Responsibilities

Researchers in the Global Flushability Lab or contract persons reporting to them will be responsible for conducting this test.

5.0 Safe Practices

This procedure involves the possible contact with septage from onsite treatment systems. Consequently, individuals performing this test may be exposed to microbiological agents that are potentially hazardous. Follow normal safety precautions when handling biologically active samples or any chemicals.

6.0 Procedure

6.1 Experimental Design

The onsite treatment test is used to assess the rate and extent of degradation of test materials in both anaerobic septic tanks and aerated onsite systems. The onsite systems used in the testing have been modified to allow for easy access into the system. This access allows for monitoring of the system and the retrieval of samples. The length of time a study runs can vary and will depend on the product being tested. The duration of the study and information regarding the pumping of the system will be included in the study records.

6.2 Concentration or Dosage Levels

The amount of product placed in each mesh bag (fiberglass screening with a 1.6 mm opening) will be described in the study protocol records.

6.3 Sample Preparation

Sample preparation will depend on the test material. The following steps explain, in general, how to prepare the samples for placement in the onsite system. Any instructions specific to the test materials will be included in the study records.

1. Fiberglass screening material (for window screens) is cut into strips approximately 6″ x 12″. Usually, eight mesh bags will be prepared for each product being tested.
2. Fold each strip of screening material to form an envelope of 6″ x 5.5″ with a 1″ flap at the top. Staple along the side edges to close the sides of the envelope to form a bag.
3. Cut approximately six 1″ strips from a durable plastic material that can be written upon with a permanent marker. Different colored plastic strips can be used for each set of samples so they may be easily identified when sampling.
4. Use the analytical balance to weigh each test material and record its weight on a table in the lab notebook.
5. Place the sample into the mesh bag, fold the flap down to close, place the plastic strip over the flap and staple along the 1″ sides and top.
6. Write the sample number and weight on the plastic strip with a Sharpie or other permanent marker (e.g., CL505-1). Each product type is identified by the sample number (–1 through –8), which corresponds to the sample weight listed in the notebook.
7. Place the sample mesh bags into a fish bag. A weight may also be needed to help keep the sample down in the tank.
8. Lower the bag into the septage and hang it by the drawstring for easy retrieval. The samples that are placed in the aerobic system should be placed in the aerated compartment.
9. On sampling days, retrieve the test sample mesh bags from the fish bag and place them in a bucket with a lid that seals.
10. Upon returning to the lab, place the bucket of samples in the freezer overnight.
11. Remove samples from the freezer the next morning and allow them to thaw. Gently rinse off the mesh bags with tap water onto a No. 18 mesh sieve (1 mm opening) using a handheld showerhead spray nozzle.

12. Carefully cut open the mesh bag and place the sample content on the No. 18 mesh sieve (1 mm opening). Gentle rinse the wastewater solids through the sieve with tap water using a handheld showerhead spray nozzle.
13. Take a picture of the sample with the digital camera and record any observations of the sample.
14. Place the sample in an aluminum cup or pan to dry in the oven at approximately 100 ± 3°C overnight or another temperature depending on the melting point of the test material. If a different temperature is required, it will be included in the study record.
15. Cool the dried sample and weigh to calculate the percent weight loss.

6.4 Type and Frequency of Tests, Analyses and Measurements

The degradation of the test material is determined by measuring the percent of weight loss over time. The sampling time points will be included in the study record or protocol. The temperature of the septage is also measured and recorded at the time of the sampling. Any other tests, analyses or measurements required will be detailed in the protocol or the study records.

6.5 Data Reporting

The data collected will be used to determine the rate and extent of degradation of the product in onsite treatment systems. A summary report will be prepared at the conclusion of the study.

6.6 Statistical Methods

Statistical methods, if used, will be described in the study records.

7.0 Documentation/Record Keeping

All records necessary to reconstruct the study and demonstrate adherence to established protocols will be maintained.

8.0 References

None

9.0 Attachments

None

A.17 Municipal Wastewater Collection and Treatment Test

1.0 Purpose

This test is used to determine the fate and transport of flushable products in wastewater collection and treatment systems.

2.0 Scope

This test evaluates the compatibility of flushable products with municipal collection and treatment systems. This test involves spiking the product in the sewer system and monitoring its fate at a wastewater treatment plant.

3.0 Principles

In this test, a flushable product is spiked into the sewer collection system and its fate at the wastewater treatment plant is monitored over time. Several locations throughout the treatment plant are monitored (e.g., bar screen, grit chamber, primary clarifier). The observed product is quantified and the percent of recovery determined. A qualitative analysis on any impact to the operation of the treatment plant is also performed.

4.0 Responsibilities

Researchers in the Flushability Lab are responsible for conducting this test. This test may be conducted by a contractor.

5.0 Safe Practices

This procedure involves the contact with raw wastewater and activated sludge from a municipal wastewater treatment plant. Consequently, individuals performing this test may be exposed to microbiological agents that are potentially hazardous. Follow normal safety precautions when handling biologically active samples or any chemicals.

6.0 Procedure

6.1 Test Design

A flushable product is either spiked into the headworks of a municipal wastewater treatment plant or spiked into the sewer collection system at a distance of approximately six hours travel time from the headworks of the municipal wastewater treatment plant. For the headworks spiking test, the fate and transport of the product is monitored throughout the treatment plant over a two-day period. Hourly observations of the unit processes are typically made over the first 12 hours and at six-hour intervals thereafter. At each sampling time point, the time, location, and condition of the product is recorded. Several locations throughout the treatment plant are monitored (e.g., headworks, bar racks, grit chamber, primary settling tanks, scum pits and wasted sludge). The bar rack scraping belt is typically turned off between observations to ensure that the product being screened is counted. For the sewer-spiking test, the product is monitored over a three-day period throughout the treatment plant. In each of the tests, anywhere from 500 to 2000 products are spiked into the system to ensure that a sufficient number can be tracked through the collection and treatment system. It is assumed that product removed by the bar rack, grit chamber or scum removal systems become solid waste and are disposed of at a landfill. Product that settles in the primary clarifier will enter the sludge digester and thus could become part of the sludge being land-applied. Product disintegration can be assessed via mesh bags in the aerobic and anaerobic digesters. Moreover, if product has been dosed over a longer period of time (e.g., during a test market study), then the biosolids and the biosolids-amended soil can be visually inspected for any evidence of the product. A soil disintegration test may be required if product is found in the digester sludge.

6.2 Spiking Test

1. Label the product with an iridescent, water insoluble marker (e.g., Sharpie) to allow for easy detection.
2. Saturate the product with water prior to discharge into the collection system. If possible, spike the product at a distance of approximately six hours of travel time from the headworks. Try to select a location at which the product will pass through at least two lift stations before reaching the treatment plant. The sewer main should be flowing at greater than ¼ capacity. The product spike load should be between 500 and 2000 products to ensure that a sufficient number can be quantified at the treatment plant.
3. Observe the fate and transport of the products through the collection system (e.g., at lift stations and manholes). Note where and whether the products are free floating or lodged. Photographs should also be taken. If access exists, collect some of the product and place it in a plastic bag for further analysis.
4. Observe the fate and transport of the products at the treatment plant over either a 48-hour or 72-hour time period. Hourly visual observation of the unit processes should be made. The time, location and condition of the products should be noted. Photographs of the product should be taken with a digital camera to visually show their condition. If access exists, collect some of the product and place in a plastic bag for further analysis.
5. Solicit wastewater treatment plant operators, on a longer-term basis, to regularly inspect unit processes and effluents for product, and note any impacts on the treatment plant operation. The longer-term observation could include routine sewer main cleaning and lift station maintenance, as well as the biosolids being amended to soil.

6.3 Data Reporting

Results from the spiking study will be summarized in a final report. This report should include product recovery and condition information. Photographs should also be included.

7.0 Documentation/Record Keeping

All records necessary to reconstruct the study and demonstrate adherence to established protocols will be maintained.

8.0 References

None

9.0 Attachments

None

WASTEWATER UTILITY

Alabama
Montgomery Water Works & Sanitary Sewer Board

Alaska
Anchorage Water & Wastewater Utility

Arizona
Gila Resources
Glendale, City of, Utilities Department
Mesa, City of
Phoenix Water Services Department
Pima County Wastewater Management

Arkansas
Little Rock Wastewater Utility

California
Calaveras County Water District
Central Contra Costa Sanitary District
Contra Costa Water District
Crestline Sanitation District
Delta Diablo Sanitation District
Dublin San Ramon Services District
East Bay Municipal Utility District
El Dorado Irrigation District
Escondido, City of
Fairfield-Suisun Sewer District
Irvine Ranch Water District
Las Virgenes Municipal Water District
Lodi, City of
Los Angeles, City of
Los Angeles County, Sanitation Districts of
Napa Sanitation District
Orange County Sanitation District
Palo Alto, City of
Riverside, City of
Sacramento Regional County Sanitation District
San Diego Metropolitan Wastewater Department, City of
San Francisco, City & County of
San Jose, City of
Santa Barbara, City of
Santa Rosa, City of
South Bayside System Authority
South Orange County Wastewater Authority
Union Sanitary District
West Valley Sanitation District

Colorado
Boulder, City of
Colorado Springs, City of
Littleton/Englewood Water Pollution Control Plant
Metro Wastewater Reclamation District, Denver

Connecticut
New Haven, City of, WPCA

District of Columbia
District of Columbia Water & Sewer Authority

Florida
Broward, County of
Fort Lauderdale, City of
Gainesville Regional Utilities
JEA
Kissimmee, City of, Department of Water Resources
Miami-Dade Water & Sewer Authority
Orange County Utilities Department
Orlando, City of
Reedy Creek Improvement District
Seminole County Environmental Services
St. Petersburg, City of
Stuart Public Utilities
Tallahassee, City of
Tampa, City of
West Palm Beach, City of

Georgia
Atlanta Department of Watershed Management
Augusta, City of
Clayton County Water Authority
Cobb County Water System
Columbus Water Works
Fulton County
Gwinnett County Department of Public Utilities
Macon Water Authority
Savannah, City of

Hawaii
Honolulu, City and County of

Idaho
Pocatello, City of

Illinois
American Bottoms Wastewater Treatment Plant
Dupage, County of
Greater Peoria Sanitary District
Kankakee River Metropolitan Agency
Metropolitan Water Reclamation District of Greater Chicago
North Shore Sanitary District
Wheaton Sanitary District

Indiana
Fort Wayne, City of

Iowa
Ames, City of
Des Moines Metro Wastewater Reclamation Authority

Kansas
Johnson County Unified Wastewater Districts
Unified Government of Wyandotte County/Kansas City, City of

Kentucky
Louisville & Jefferson County Metropolitan Sewer District

Louisiana
Sewerage & Water Board of New Orleans

Maine
Bangor, City of
Portland Water District

Maryland
Anne Arundel County Bureau of Utility Operations
Howard County Department of Public Works
Washington Suburban Sanitary Commission

Massachusetts
Boston Water & Sewer Commission
Upper Blackstone Water Pollution Abatement District

Michigan
Ann Arbor, City of
Detroit, City of
Grand Rapids, City of
Holland Board of Public Works
Lansing, City of
Owosso Mid-Shiawassee County WWTP
Saginaw, City of
Wayne County Department of Environment
Wyoming, City of

Minnesota
Rochester, City of

Missouri
Independence, City of
Kansas City Missouri Water Services Department
Little Blue Valley Sewer District
Metropolitan St. Louis Sewer District

Nebraska
Lincoln Wastewater System

Nevada
Henderson, City of

New Jersey
Passaic Valley Sewerage Commissioners

New York
New York City Department of Environmental Protection

North Carolina
Cary, Town of
Charlotte/Mecklenburg Utilities
Durham, City of
Metropolitan Sewerage District of Buncombe County
Orange Water & Sewer Authority

Ohio
Akron, City of
Butler County Department of Environmental Services
Columbus, City of
Metropolitan Sewer District of Greater Cincinnati
Northeast Ohio Regional Sewer District
Summit, County of

Oklahoma
Tulsa, City of

Oregon
Clean Water Services
Eugene/Springfield Water Pollution Control
Water Environment Services

Pennsylvania
Philadelphia, City of
University Area Joint Authority, State College

South Carolina
Charleston Commissioners of Public Works
Mount Pleasant Waterworks & Sewer Commission
Spartanburg Sanitary Sewer District

Tennessee
Cleveland, City of
Knoxville Utilities Board
Murfreesboro Water & Sewer Department
Nashville Metro Water Services

Texas
Amarillo, City of
Austin, City of
Dallas Water Utilities
Denton, City of
El Paso Water Utilities
Fort Worth, City of
Gulf Coast Waste Disposal Authority
Houston, City of
San Antonio Water System
Trinity River Authority

Utah
Salt Lake City Corporation

Virginia
Alexandria Sanitation Authority
Arlington, County of
Fairfax County Virginia
Hampton Roads Sanitation District
Henrico, County of
Hopewell Regional Wastewater Treatment Facility
Loudoun County Sanitation Authority
Lynchburg Regional WWTP
Prince William County Service Authority
Richmond, City of

Washington
Edmonds, City of
Everett, City of
King County Department of Natural Resources
Seattle Public Utilities
Yakima, City of

Wisconsin
Green Bay Metro Sewerage District
Madison Metropolitan Sewerage District
Milwaukee Metropolitan Sewerage District
Racine, City of
Sheboygan Regional Wastewater Treatment
Wausau Water Works

Australia
South Australian Water Corp.
Sydney Water Corp.
Water Corp. of Western Australia

Canada
Greater Vancouver Regional District
Toronto, City of, Ontario

Mexico
Servicios de Agua y Drenaje de Monterrey, I.P.D.

United Kingdom
Yorkshire Water Services Limited

STORMWATER UTLITY

California
Los Angeles, City of, Department of Public Works
Monterey, City of
Santa Rosa, City of

Colorado
Boulder, City of

Georgia
Griffin, City of

Kansas
Overland Park, City of

Kentucky
Louisville & Jefferson County Metropolitan Sewer District

Maine
Portland Water District

North Carolina
Cary, Town of
Charlotte, City of, Stormwater Services

Pennsylvania
Philadelphia, City of

STATE

Fresno Metropolitan Flood Control District, Calif.
Urban Drainage & Flood Control District, Colo.

CORPORATE

ADS Environmental Services
The ADVENT Group Inc.
Alan Plummer & Associates
Alden Research Laboratory
Alpine Technology Inc.
Aquateam–Norwegian Water Technology Centre A/S
BaySaver Inc.
BioVir Inc.
Black & Veatch
Boyle Engineering Corporation
BPR CSO
Brown & Caldwell
Burns & McDonnell
CABE Associates Inc.
The Cadmus Group
Camp Dresser & McKee Inc.
Carollo Engineers Inc.
Carpenter Environmental Associates Inc.
CDS Technologies Inc.
Chemtrac Systems Inc.
CH2M HILL
Clancy Environmental Consultants Inc.
Damon S. Williams Associates, LLC
David L. Sheridan, P.C.
Earth Tech Inc.
Ecolab Water Care Services
EMA Inc.
The Eshelman Company Inc.
Finkbeiner, Pettis, & Strout Inc.
Frontier Geosciences Inc.
ftn Associates Inc.
Gannett Fleming Inc.
GE Betz
Golder Associates Inc.
Greeley and Hansen LLC
The HACH Company
Hazen & Sawyer, P.C.
HDR Engineering Inc.
HNTB Corporation
HydroQual Inc.
Insituform Technologies Inc.
Institute for Environmental Technology & Industry, Korea
Jacobson Helgoth Consultants Inc.
Jason Consultants Inc.
Jordan, Jones, & Goulding Inc.
KCI Technologies Inc.
Kelly & Weaver, P.C.
Kennedy/Jenks Consultants
Komline Sanderson Engineering Corporation
Lawler, Matusky & Skelly Engineers, LLP
Limno-Tech Inc.
Lombardo Associates Inc.
Malcolm Pirnie Inc.
McKim & Creed
MEC Analytical Systems Inc.
MWH
New England Organics
Odor & Corrosion Technology Consultants Inc. (OCTC)
ONDEO Degremont Inc.
Oswald Green, LLC
PA Government Services Inc.
Parametrix Inc.
Parsons
Post, Buckley, Schuh & Jernigan
R&D Engineering/Conestoga Rover & Associates
The RETEC Group
R.M. Towill Corporation
Ross & Associates Ltd.
Royce Technologies
SAIC Maritime Technical Group
Sear-Brown
Sierra Environmental Services
SYNAGRO
Tetra Tech Inc.
Trojan Technologies Inc.
URS Corporation
USFilter NATC
Vortechnics Inc.
Wade-Trim Inc.
Weston Solutions Inc.
Woodard & Curran
WRc/D&B, LLC
WWETCO, LLC

INDUSTRY

American Electric Power
ChevronTexaco Energy Research & Technology Company
The Coca-Cola Company
Dow Chemical Company
DuPont Company
Eastman Kodak Company
Merck & Company Inc.
ONDEO Services
Procter & Gamble Company
PSEG Services Corp.
Severn Trent Services Inc.
Shell Global Solutions
Thames Water Plc
United Water Services LLC

Note: List as of 5/12/03

Board of Directors

Chair
Stephen T. Hayashi
Union Sanitary District

Vice-Chair
James F. Stahl
County Sanitation Districts of Los Angeles County

Secretary
William J. Bertera
Water Environment Federation

Treasurer
Karl W. Mueldener
Kansas Department of Health & Environment

Robert Berger
East Bay Municipal Utility District

Gordon R. Garner
Louisville & Jefferson County Metropolitan Sewer District

Lawrence P. Jaworski
Greeley & Hansen

Richard D. Kuchenrither, Ph.D.
Black & Veatch Corporation

Vernon D. Lucy
ONDEO Degremont Inc.

Robert T. McMillon
Fort Worth Water Department

Thomas R. Morgan
Montgomery Water Works & Sanitary Sewer Board

Philip C. Singer, Ph.D.
University of North Carolina, Chapel Hill

Scott M. Summers
Eastman Kodak Company

Murli Tolaney
MWH

Donnie R. Wheeler
Hampton Roads Sanitation District

Executive Director
Glenn Reinhardt

Research Council

Chair
Robert Berger
East Bay Municipal Utility District

Vice-Chair
John Thomas Novak, Ph.D.
Virginia Polytechnic Institute & State University

Robert G. Arnold, Ph.D.
University of Arizona, Tucson

Robin L. Autenrieth, Ph.D.
Texas A&M University

James Crook, Ph.D.
CH2M HILL

James R. Dartez
Royce Instrument Corporation

Philip B. Dorn, Ph.D.
Equilon, LLC

Geoffrey H. Grubbs
U.S. Environmental Protection Agency

Michael D. Jawson, Ph.D.
U.S. Department of Agriculture

Norman E. LeBlanc
Hampton Roads Sanitation District

Margaret H. Nellor, P.E.
County Sanitation Districts of Los Angeles County

Spyros Pavlostathis, Ph.D.
Georgia Institute of Technology

Robert A. Reich
DuPont Engineering Company

Tyler Richards
Gwinnett County

Prakasam Tata, Ph.D.
Metropolitan Water Reclamation District of Greater Chicago

George Tchobanoglous, Ph.D.
Tchobanoglous Consulting

Gary Toranzos, Ph.D.
University of Puerto Rico

Stormwater Technical Advisory Committee

Chair
Christine Andersen, P.E.
City of Boulder

Gail B. Boyd
URS Corporation

Larry Coffman
Prince George's County

Doug Harrison
Fresno Metro Flood Control District

Robert E. Pitt, Ph.D., P.E., D.E.E.
University of Alabama

A. Charles Rowney, Ph.D.
Camp Dresser & McKee Inc.

Ben Urbonas, P.E.
Urban Drainage and Flood Control District

James Wheeler, P.E.
U.S. Environmental Protection Agency

WERF Product Order Form

a benefit of joining the Water Environment Research Foundation, subscribers are entitled to receive one complimentary copy of all final orts and other products. Additional copies are available at cost (usually $10). To order your complimentary copy of a report, please write e" in the unit price column. WERF keeps track of all orders. If the charge differs from what is shown here, we will call to confirm the total ore processing.

me Title

ganization

dress

y State Zip Code Country

one Fax Email

Stock #	Product	Quantity	Unit Price	Total

Postage & Handling	
VA Residents Add 4.5% Sales Tax	
Canadian Residents Add 7% GST	
TOTAL	

Method of Payment:

(All orders must be prepaid.)

- Check or Money Order Enclosed
- Visa ❑ Mastercard ❑ American Express

ccount No. Exp. Date

gnature

Shipping & Handling:

Amount of Order	United States	Canada & Mexico	All Others
Up to but not more than:	Add:	Add:	Add:
$20.00	$5.00*	$8.00	50% of amount
30.00	5.50	8.00	40% of amount
40.00	6.00	8.00	
50.00	6.50	14.00	
60.00	7.00	14.00	
80.00	8.00	14.00	
100.00	10.00	21.00	
150.00	12.50	28.00	
200.00	15.00	35.00	
More than $200.00	Add 20% of order	Add 20% of order	

*minimum amount for all orders

Note: Please make checks payable to the Water Environment Research Foundation.

To Order (Subscribers Only):

Log on to www.werf.org and click on the "Product Catalog."

Phone: (703) 684-2470
Fax: (703) 299-0742.

WERF
Attn: Subscriber Services
601 Wythe Street
Alexandria, VA 22314-1994

To Order (Non-Subscribers):

Non-subscribers may be able to order WERF publications either through WEF (www.wef.org) or IWAP (www.iwapublishing.com). Visit WERF's website at www.werf.org for details.

IWA Publishing's authorised EU representative for General Product Safety Regulations is Diane D'Arras, 15 rue Duret, 75116 Paris, France, e-mail: safety@iwap.co.uk.

Printed and bound by CPI Group (UK) Ltd, Croydon, CR0 4YY
12/05/2026
02108879-0011